MANAGEMENT LAUREATES:

A Collection of Autobiographical Essays

VOLUME 4

Editor: ARTHUR G. BEDEIAN
Ralph and Kacoo Olinde Distinguished Professor
College of Business Administration, Louisiana State University

MANAGEMENT LAUREATES:

A Collection of Autobiographical Essays

by

KATHRYN M. BARTOL
JANICE M. BEYER
GEERT HOFSTEDE
JOHN M. IVANCEVICH
FRED LUTHANS
JEFFREY PFEFFER
DEREK S. PUGH
JOHN W. SLOCUM, JR.

JAI PRESS INC.

Greenwich, Connecticut *London, England*

55 Old Post Road No. 2
Greenwich, Connecticut 06830

JAI PRESS LTD.
38 Tavistock Street
Covent Garden
London WC2E 7PB
England

ISBN: 1-55938-730-0

Manufactured in the United States of America

CONTENTS

PREFACE

The initial volumes of *Management Laureates: A Collection of Autobiographical Essays* (Bedeian, 1992-1993) were the outcome of many years of contemplation. They were prompted by the belief that it is difficult to fully understand the work of the management discipline's leading thinkers without knowing more about them as individuals. The goal of the present volume, therefore, is identical to that of its predecessors: to present autobiographies of the discipline's most distinguished laureates.

Prior to the *Management Laureates* series, the literature focused almost exclusively on methodology or "what one does," as authentication for one's findings. Largely ignored was the impact of the motivation and thoughts that define "who one is" (cf. Lederberg, 1990). The problem with this omission is that all scholarship is autobiographical. There is no theory, no hypothesis, no proposition that is not a fragment of some autobiography (Söderqvist, 1991). Recognizing the "autobiographical character of knowledge construction," McCloskey (1986) argues, and I agree, that autobiography is often the best way to understand someone's point of view. If nothing else, autobiography does much to unmask the purported objectivity of methodized science and explain why academic presentations so frequently began with personal explanations of "how I came to this problem" (McCloskey, 1985, p. xi).

Because the relation between management as a scientific discipline and the forces that shaped the lives of those associated with this discipline had seldom been considered, management literature lacked all but the faintest clues as to the personal and social conditions that had informed the discipline's substantive development. The autobiographies in this and other volumes of the *Management Laureate* series remedy that situation. They offer the reader not only a glimpse of the background and personal experiences of the discipline's most distinguished laureates—their frustrations, triumphs, and transgressions—but also a deeper understanding of what management is and what it is becoming.

The narratives in the *Management Laureates* series establish connections between the life experiences of the contributing laureates and the development of their scholarship. These connections, in turn, further illuminate the management discipline. By clarifying the extent to which the private and professional worlds of their authors operate *across* rather than *within* separate spheres, the autobiographies in this series highlight that not one of the authors has worked or thought in a vacuum (Harrison & Lyon, 1993). Each has functioned within an intellectual milieu of ideas and people. For this reason, the following accounts are, more accurately, an *admixture* of autobiography (wherein attention is focused on the self) and memoir (wherein attention is focused on others; Pascal, 1960). This combination reaffirms Stanley's (1993) observation that "a life, whether of one's self or another, is never composed of one decorticated person alone" (p. 50). In the testaments that follow, the reader will see that this is true for each of the contributing laureates. Each has moved among a variety of overlapping intellectual and social networks, internalizing aspects of these networks, and, at the same time, leaving indelible marks on the institutions and people in these networks.

Clearly then, each autobiography contains various "layers" of information (Szczepański, 1981). Bear in mind, however, that no autobiography—regardless of the number of layers—is a *true* picture of the author's life. It is, instead, what he or she wishes (or is impelled) to reveal at a particular moment. Autobiographies, therefore, reflect a "special truth" colored by both conscious and unconscious omissions, as well as inevitable distortions and illusions (Kendall, 1965; p. 30). "Forced to confront their own lives," authors may well "mute their passions, avoid vivisection of their motives, portray themselves as bemused players in a game they do not run, [and] treat their own careers as a series of breaks, lucky or otherwise" (Tilly, 1993, pp. 497-498).

The extent to which the contributing laureates have or have not offered graphic glimpses of their lives and, possibly, experienced related attacks of "prudence, modesty, and propriety" is unknown. It is safe to say, however, that their autobiographies, like all others, are a "balance between the self and the world, the subjective and the objective," casting light upon both (Pascal, 1960, p. 181).

The manner in which autobiographers achieve this balance between the self and the world is, in itself, insightful. In selecting facts, incidents, and events for presentation, they necessarily reveal something about their own attitudes, values, and beliefs (Burnett, 1974/1977). As MacLure (1993) explains, one's representation of one's personal life history is a means by which one justifies, explains, and makes sense of oneself in relation to others. The value of such self-portrayals lies in their status as "recollected experience rather than recorded fact" (Scott, 1990, p. 177). Beyond doubt, as Burnett (1982/1984) argues, "The autobiographer's version of what happened to himself possesses a personal validity which is different in kind from any second-hand account, however skilled the reporter may be in techniques of observation and analysis .... The very partiality of the account is, therefore, part of its value for the author has chosen his own ground, patterned his own experiences, and has painted a self-portrait which is more revealing than any photograph" (p. 11).

The autobiographies included in the present volume do not shed light, obviously, on the entire management discipline. They do, however, provide a window into what goes on in the lives of the discipline's leading thinkers. As long-standing "members of the guild," the laureates in this volume have each made significant contributions to the management enterprise. All possess a remarkable strength of character and resiliency. As their autobiographies reveal, however, each is singular in his or her definition of fulfillment and the specific configuration of choices enacted in his or her life's journey. Their willingness to share their thoughts and experiences with generations of future readers is most appreciated. It is hoped that each laureate will take pride and satisfaction in the immortality that the present undertaking provides and in the knowledge that he or she has hereby gained a medium of access that will allow each, decades from now, to speak directly, and personally, to those management scholars who will be heirs to their intellectual legacy.

Arthur G. Bedeian
November 1995

REFERENCES

Bedeian, A.G. (Ed.). (1992-1993). *Management laureates: A collection of autobiographical essays* (Vols. 1-3). Greenwich, CT: JAI Press.

Burnett, J. (1974/1977). Autobiographies as history. In J. Burnett (Ed.), *Useful toil: Autobiographies of working people from the 1820s to the 1920s* (pp. 9-19). Harmondsworth, Middlesex, ENG: Penguin.

Burnett, J. (1982/1984). Autobiographies as history. In J. Burnett (Ed.), *Destiny obscure: Autobiographies of childhood, education and family from the 1820s to the 1920s* (pp. 9-17). Harmondsworth, Middlesex, ENG: Penguin.

Harrison, B., & Lyon, E. S. (1993). A note on ethical issues in the use of autobiography in sociological research. *Sociology, 27*, 101-109.

Kendall, P.M. (1965). *The art of biography.* New York: Norton.

Lederberg, J. (1990). Introduction. In J. Lederberg (Ed.), *The excitement and fascination of science: Reflections by eminent scientists* (Vol. 3, pp. xvii-xxiv). Palo Alto, CA: Annual Reviews.

MacLure, M. (1993). Arguing for your self: Identity as an organising principle in teachers' jobs and lives. *British Educational Research Journal, 19*, 311-322.

McCloskey, D.N. (1985). *The rhetoric of economics.* Madison: University of Wisconsin Press.

Pascal, R. (1960). *Design and truth in autobiography.* London: Routledge & Kegan Paul.

Söderqvist, T. (1991). Biography or ethnobiography or both? Embodied reflexivity and the deconstruction of knowledge-power. In F. Steier (Ed.), *Research and reflexivity* (pp. 143-162). Beverly Hills, CA: Sage.

Stanley, L. (1993). On auto/biography in sociology. *Sociology, 27*, 41-52.

Szczepański, J. (1981). The use of autobiographies in historical social psychology. In D. Bertaux (Ed.), *Biography and society: The life history approach in the social sciences* (p. 225-234). Beverly Hills, CA: Sage.

Tilly, C. (1993). Blanding in. [Review of the book *Authors of their own lives: Intellectual autobiographies by twenty American sociologists*]. *Sociological Forum, 8*, 497-505.

If I have seen a little farther it
is by standing on the shoulders of Giants.

—Sir Issac Newton
5 February 1676

Kathryn M. Bartol

Challenged on the Cutting Edge

KATHRYN M. BARTOL

When I was grade-school aged in Philadelphia, I remember hearing the bible story of the talents. As the story goes, two servants were each given a certain number of talents (coins). One used the talents well and multiplied the master's investment many times over. The other buried the talents to keep them safe—of course, squandering the opportunity. The story impressed me. I could see the applicable logic that everyone is born with a certain set of natural abilities. Individuals can then either try to develop their abilities and use them well or can leave much of their potential unrealized. At any rate, I evolved the hope that I could develop my talents over the course of my life and use them to make differences in various ways.

My maternal grandfather had been an engineer for Baldwin Locomotive, a Philadelphia-based maker of steam engines. A recent article in *Forbes* magazine noted that Baldwin Locomotive was one of the top 100 companies in 1917, when *Forbes* began publishing. When the technological innovation of the diesel engine displaced the steam engine, Baldwin did not adapt quickly enough. My grandfather was displaced from his job, left the company and began a new career in insurance. He would sometimes talk about his experiences at Baldwin, noting in particular that Baldwin managers tended to take the attitude that "there will always be a need for steam engines." (Several years ago, my husband and I took a trip from Alexandria, Virginia, to Charlottesville, Virginia, on a train pulled by a steam engine. The engine was

Management Laureates, Volume 4, pages 1-37.

ISBN: 1-55938-730-0

a marvelous piece of technology. However, the train had to stop periodically for coal and water for the engine, adding considerably to travel time. Moreover, we learned that if you opened the window in cars that were sufficiently down wind from the engine's plume of smoke, you became covered with soot!) I always found the story intriguing, particularly management's role in the company's failure to recognize and cope with the cutting edge of change. An immediate implication was the situation caused my grandfather to decide that a college education was very important and would be more so in the future. He was forever clipping articles out of newspapers and magazines that discussed the value of going to college. The information campaign was mainly aimed at my older brother because women generally did not have long-term careers at the time, but much of it rubbed off on me. To my grandfather's credit, he was fairly supportive of my own interest in attending college and having a career. When my brother wanted to give up his weekly paper route and I wanted to take it over, my grandfather dutifully helped me deliver the papers for several weeks until I finally concluded that maybe a paper route for a 9-year-old girl was too much of a burden for all concerned. Fortunately he lived a long life and saw my brother receive both an undergraduate and a graduate degree and me graduate from college.

My father, Walter R. Ottinger, was a government auditor and traveled about three-quarters of the time. My mother, Mary Scherf Ottinger, helped out with expenses by typing up depositions and other court documents from steno typed notes. This way she could work mainly at home. As I advanced to the upper grades of grade school, she began taking more depositions herself and I became an early model of the latch-key kid. I very much admired my mother for working, although I gradually became aware that she was violating sex-role norms. People would occasionally make nasty remarks to the effect that she was neglecting her children—mainly meaning me, because my brother was several years older. Sometimes after one of these incidents, my mother would feel bad and ask me if I felt neglected. To the contrary, I would assure her that I was proud of her. The situation gave me early insight in some of the difficulties caused by gender roles and the challenge of changing them. For my part, I always pictured myself being a part of the world of work.

My mother learned to be a court reporter as a student at Strayer's Business College in Philadelphia, considered at the time to be a premier training school for secretarial and related business careers. My mother did so well there that she was asked to stay as an instructor after she graduated. In this post-depression era, money was scarce and my father completed Stayer's with free tuition as part of my mother's pay. The Strayer's experience meant that both of my parents had excellent writing skills. My mother had won some type of writing contest in high school and always encouraged me to write. She and my father were good sounding boards for my grade school and high school essays. As a result, I developed an appreciation for writing skills, although I loved math as well.

To cut the time away from home, my father accepted a position working for the U.S. Customs Service in Detroit, later becoming deputy director of what was then the United States's second busiest port. We moved to the Detroit area when I was a sophomore in high school. Although I did extremely well in high school from a grade point-of-view, in retrospect the environment did not challenge me enough.

When I became a student at Marygrove College in Detroit, which was a Catholic, all-female, liberal arts college at the time, I lived on campus. During that era, we had limited off-campus privileges on weekday nights, making the atmosphere very conducive to studying. Moreover, the training was rigorous. The English Department, in particular, took great pride in its ability to train students to do high-quality library research and write up the results in a clear, logical way. Performance expectations were high. The faculty claimed that their approach was excellent preparation for graduate school. In my ensuing years of graduate school I found their claims to be well founded. More importantly, the training left me with a deep appreciation of and an interest in doing research.

When I started college, I had planned to major in mathematics. Although I could perform the work in the math classes quite well, I found it difficult to fathom the big picture—i.e., why what we were doing mattered. Also it was unclear to me at the time what type of career path was available through mathematics. Mostly, I heard programming mentioned. Ironically, through a completely different route, I would end up working in the computer field and also ultimately acquire a strong statistical background. But to me ending my freshman year, programming sounded very boring and I switched to majoring in journalism.

As part of the journalism program, the journalism students produced the school paper that came out bi-weekly. I was fortunate to be assigned as the reporter for the president of the College, Sr. Honora Jack, I.H.M. Over the next three years, I gained valuable insight into the problems of running a sizable organization. Sr. Honora was gracious, highly-educated, charismatic, and visionary. One of her vision was that graduates would take leadership roles in working to make the world a better place, particularly helping individuals less fortunate than they. Not surprisingly, then, graduate school was very much valued. The atmosphere fostered a vision of women as capable of being successful in any field. Although the possibility of discrimination in some fields was acknowledged, it was not generally perceived as an insurmountable obstacle.

During the summer following my sophomore year, I took an American history course in the evening at the University of Detroit. The course was taught by Robert (Bob) A. Bartol, who was doing graduate work in history. As I saw it at the time, he had the annoying habit of calling on class members individually to answer detailed questions about the chapter that we were supposed to have read. As a result, we had to know the material exceptionally well for each class. Because I was working in the mortgage department of Northwestern Mutual Life Insurance Company for my second summer, I had a very difficult time keeping up with all the reading and knowing it that well for each class. Also, it seemed that the instructor was particularly tough on me as we steamed in the heat of the classroom which lacked air-conditioning. Although I learned a great deal, I was very happy when the class was over, never realizing what impact that class would have on my life. About a year later, I ran into Bob at a social function and we began dating. I found his extensive knowledge of the history of many parts of the world to be fascinating.

While I studied journalism, I became particularly interested in advertising. The summer preceding my senior year, I was fortunate to be one of 50 college seniors across the country selected to attend a College Advertising Week sponsored by the Advertising Club of New York. The week consisted of a number of presentations at major advertising agencies in New York and discussions with some of the top advertising personalities. One of the valuable pieces of information I learned during the week was that it was virtually impossible to obtain a writing job with a New York agency right out of college. Instead, one had to accept a position in the mailroom or as a secretary and hope for an eventual chance to do some writing.

The other way into the advertising field was to take a job doing retail advertising, build a book of tear sheets (copies of one's published ads), and then attempt to obtain a copywriter position with an ad agency. Either way, entry level jobs in the advertising field did not pay well. Going to New York and hoping for a chance to write seemed to be an extremely risky path that could burn up a great deal of valuable time. To test the situation, I tried a number of advertising agencies around the Detroit area and was commonly told, "Sorry, we don't hire women for those positions." One notable exception was the major advertising agency, Campbell-Ewald, which did have me put together some hypothetical sample ads and also interviewed me.

Meanwhile, I put a great deal of effort into applying for retail advertising positions. After a long year and a major campaign, including putting together several ads as part of a "creativity test," I was very fortune in landing the job I really wanted. I was hired as a copywriter with the J. L. Hudson Company, the major department store chain in the Detroit area, which is now part of the Dayton-Hudson chain. Any illusions of advertising glamour were at least temporarily shattered by my first assignment in which I was allocated a maximum of 30 words to describe to virtues of a trash can! During the year that I was there, the advertising department won the Retail Advertising Award of the year (certainly not due to my neophyte efforts!). The people that I worked with were consummate professionals and I learned a great deal, including how hectic getting advertising to the right place and at the right time could be. Out of necessity, I learned to compose on the typewriter, to work fast, and to juggle many projects at once. Because we worked closely with buyers in different departments, we actually obtained considerable insight into how a major department store runs (including strategy, competitive analysis, purchasing, delivery problems, communication problems, internal politics, and the impact of different types of pay plans on the sales force).

Bob Bartol had gone on to the University of Michigan for further graduate work. He urged me to go to graduate school at Michigan. Although at Hudson's I was rather quickly moved into some of the most interesting areas for which to write, such as fine furniture and fashions, I began to notice that there was a cycle to things—e.g., Mother's day, Father's day. I could envision myself eventually becoming bored, despite the changing fashions and the tremendously creative people that I worked with.

After a few trips to Ann Arbor, I applied to graduate school at Michigan and was accepted for the following fall. A few weeks later, Campbell-Ewald, the advertising agency, called and offered me a copywriter job, but the die was cast for Michigan. Bob had helped me line up a dorm counselor job, which would pay for my room and board, but when I got to Ann Arbor, the Journalism department pressured me into also accepting an assistantship working in the department office. So I was extremely busy with two jobs and spent every possible spare moment studying. Although the students were fun, the dorm job was exasperating because there were many unforeseen interruptions (especially when I had an exam the next day, it seemed). Fortunately the following year I was able to increase my hours in the Journalism department and quit the dorm job. Bob and I loved the graduate library at Michigan because it had so many resources. We spent many hours there, particularly on weekend nights when things were relatively quiet and we could access materials very quickly.

The master's degree program was a two-year program in which a considerable portion of the work was in a concentration outside of the Journalism department. Bob was majoring in Russian History and I became interested in it also. As a result I made my concentration Russian Studies, necessitating taking two years of Russian language training and graduate courses in a variety of departments, such as history, literature, and geography. I was often thrown in with doctoral students from those departments, but managed to hold my own. One particularly excellent teacher who influenced me greatly was Dr. Alfred Levin, a Russian historian who was visiting from the University of Oklahoma. Based on my work in his doctoral seminar, he urged me to obtain a Ph.D. in history. More importantly, interviewing him for a newspaper published by the journalism department did plant the seed in my mind that maybe it could possibly be more interesting to be a subject matter expert—i.e., the interviewee—rather than the interviewer. Still, being a journalist did involve a great deal of research, particularly if one wanted to add value and do a first-class job.

Because the master's level was the terminal degree in the Journalism department, the master's students were the focal point of the faculty. There were seminars in which nationally-known journalists were often flown in as speakers. The major seminar was held at the chairperson's home and was informal enough that we could ask many questions of the guest speakers. Many of the award-winning journalists, such as David Halberstam, were fascinating and we gained great insight into how top journalists operated. It was exciting, but at times unsettling. Mainly, I found myself dissatisfied with the process that some (though certainly not all) of the top journalists used to learn about new areas in which they were reporting. I asked one famous journalist who was reporting for the *New York Times* about how he prepared when he had to go to a new country to report the news. He told us that he usually read *one* book! Concentrating in Russian Studies had helped me understand how difficult it is to understand another culture. I found his answer frightening.

Bob decided to do his doctoral work with Dr. Arthur Adams, a prominent Russian Historian who was teaching at Michigan State University in East Lansing at the time. We were soon to be married and I was nearing the end of my master's program, so I began to concentrate my job efforts on the Lansing area. Naturally I applied for a job with the major newspaper in Lansing. Because Lansing was the state capital and there were not many journalism opportunities in the relatively small city, I also took the Civil Service Exam for a job in state government—just in case.

However, it turned out that if you did not interview for jobs as they became available in your area of expertise, you would be removed from the Civil Service applicant list. So I began to go to interviews in such areas as government relations, systems analysis, and related professional classifications. Although interviews with three government agencies had gone well, I sensed a reluctance to hire me. An interview with a fourth state agency, the Michigan Department of Social Services, however, had gone particularly well.

Meanwhile, I had been interviewed by the major Lansing newspaper. On my second interview with the newspaper, the editor told me that he wanted to hire me. He then went on at length about the merit pay system that I would be working under, which would allow me to earn extra pay for very good work. He also emphasized that he did not want me to discuss my pay arrangement with the other employees. Suddenly, he looked at my left hand, noticed the engagement ring (which had been there during the first interview as well), and exclaimed, "Oh, your getting married? Well, I'll still give you the job, but forget everything I said about merit pay." He gave me a deadline of two weeks to accept the job. Naturally, I was upset over the editor's reaction to my change in marital status. Also I was troubled by the fact that he was so concerned that I not discuss with other employees the merit pay arrangement he had outlined. Although I realized that developing fair pay systems is difficult, the incident created mistrust in my mind without even considering the sexist attitude issue.

As I pondered the situation, I received a request for a second interview at the Michigan Department of Social Services. The job was a systems analyst position. My job would be to study information flows and figure out ways to make the Department's systems operate more efficiently. Because so many of the issues involved communication of one sort or another, the managers thought that I would bring a fresh perspective to the situation. They offered me the job. It was not exactly what I had had in mind, but it sounded interesting and the pay was surprisingly good. After the pay discussion with the newspaper, I felt I would be exploited there. So I took the position with the Department of Social Services. It was a fateful decision, because I had actually taken a major step toward a career in the field of management.

The Department was interesting from a gender distribution point of view. There were many professional women working as social workers in the various counties, but only one woman in a top position at the state level and she resigned shortly after I began working there. Though relatively low level, I was actually one of very few professional women working in the state office (There were very few professional women working at the state level in the other state agencies, as well.) Under the circumstances, I was very lucky to have obtained the job and continue to be grateful to the people who hired me.

A month after I started work, top management decided the Department should increase its use of computers and more people would need to become knowledgeable about computers. I quickly volunteered to take an IBM aptitude test and was one of several individuals subsequently chosen to attend a series of training programs aimed at making us computer literate. The ultimate goal was to streamline and further computerize the paperwork process that distributed payments and other services to thousands of welfare recipients in the State of Michigan.

The situation turned out to be a change-management challenge par excellence. There were several interesting aspects to the problem. For one thing, to reengineer the paperwork processes, it was necessary to understand the processes very well. However, clerical staff members involved in the paperwork process tended to be fearful of the impact of the computer on their jobs. As we attempted to chart the paperwork flow to understand it thoroughly, we sometimes met with considerable resistance. I spent many hours working with staff members attempting to gain their trust so that ultimately I was able to obtain the needed information.

A second challenge was that many of the social workers tended to take the view that their job was to help people. Paperwork of any kind was a plague to be avoided as much as possible. As a result, they often completed the paperwork incorrectly, ironically causing their client not to receive a welfare check or other important service! We attempted to gain their cooperation by making the computerized system as user friendly as possible. But it was usually a hard sell, and I spent a fair amount of time visiting Wayne County (the Detroit area), where the bulk of Michigan's population and our caseload resided.

A third challenge was that I was often in the middle between the social workers who wanted an interface that was technically impossible to achieve and some of the technical people who wanted people to conform to what could be most easily programmed. In those days, computers were not terribly flexible. I spent a great deal of time learning both the computer end and the complex requirements of various welfare programs so that I could communicate with both sides in brokering the interface.

Congress had just passed the Medicaid program, which states were to implement. The regulations were complicated, confusing, and incomplete. One purpose of the program was to pay for the health care of individuals receiving welfare. Another was to pay for the nursing home care of elderly people who had exhausted their own resources and required nursing home care. The system involving nursing home care was particularly complex, because it was necessary to match eligibility from the records provided by caseworkers with information received from nursing homes. Yet the population was quite mobile in the sense of going in and out of hospitals and moving from one nursing home to another for various reasons. Moreover, eligibility requirements were somewhat murky and changes often had to be made, sometimes retroactively. Originally, an outside vendor was handling the payments, but the system was quite costly and a state audit unearthed many problems with payments made (duplicate payments, payments for ineligible people, delayed payments, etc.). The situation was serious because the federal government would not provide funding to the State for payments made in behalf of ineligible people. In addition the nursing home owners were complaining vociferously to their state legislators about delayed payments. Ultimately, the State decided to take over the system that dispensed more than $300-million in payments annually and I was asked to head up the project.

Doing this job gave me valuable first-hand experience regarding the difficulties of operating as a matrix manager. Although I was responsible for completing the project on schedule and within budget, the systems analysts and programmers who were assigned to work for me also reported to their respective functional managers. The project also required considerable cooperation and help from data entry, computer operations, the accounting department, and the clerical processing unit. Because so many operations were being computerized, the entire computer section was under great work pressure and resources were tight. As we began our work, I noticed that the programmers were putting little time and effort into the project. After talking with the lead programmer who was usually very conscientious, I learned that he had been on another major project like this earlier, which had failed. He said the programmers had been blamed even though they believed that many of the problems were not their fault. As a result, they had little enthusiasm for working on major projects like this one.

After considerable thought, I called a meeting of the various programmers assigned to the project. I explained with great confidence (probably partially attributable to my naivete) that this project had high visibility and was definitely going to be a success. I noted that each programmer had been carefully chosen to be a part of this project. On the other hand, I said, things had to change. This was the time for each person to decide whether they were on board or wanted off (there was heavy risk in this strategy because the available options were much less capable). The programmers at the meeting always claimed that I pounded the table when I talked about getting on with the project or getting off (I didn't remember doing that) and kidded me about it for the duration of the project. After my assurances that we would be successful and that I would make sure that they were not blamed for anything that wasn't within their control, they signed on and did a fantastic job.

Over the next 1½ years, we worked at a grueling pace. The federal regulations associated with Medicaid and with the welfare system in general were not designed from a management point-of-view at all. Many of the required procedures relating to checking and verifying client eligibly were costly, time-consuming, and a disaster from a cost-benefit perspective. Thus, I gained valuable insight into many of the reasons why government often is bureaucratic and bogged down in red tape. Because of my need to understand the regulations well to oversee the system design, I studied them closely and posed a number of questions for which policy makers obtained answers from Washington. I gradually became perceived as one of the most knowledgeable persons in the State regarding the requirements and, as a result, I began to be included in high-level policy meetings regarding the Medicaid program itself.

Our project team consisted of a core group of two systems analysts and eight programmers, though we often used more programmers during critical times. Because the contract with the vendor currently paying bills had a specific ending date, it was imperative that the project be completed on time to take over on that date. In addition, complaints by nursing home operators to the legislature made this a project with considerable downside risk—a failure would be very visible. To try to find a way to control this large and complex project which had to be completed on time, I did library research and discovered PERT. To my delight, I also found that a PERT program was available from IBM, our computer vendor, and I spent many hours developing a PERT diagram, estimating completion times, etc. The time proved to be well worth it, because it allowed me to keep track of a large number of diverse pieces that could be worked on simultaneously. We also had the graphics people put together a color coded version of the diagram, which upper-level management loved to the point of periodically taking it over to the state legislature to report progress.

As is typically the case with project-oriented matrix management, the functional area managers often wanted to "borrow" key systems analysts and programmers from the project to handle immediate pressing problems.

Because the systems analysts and programmers ultimately worked for their functional bosses and would return to their function department after the project, they naturally wanted to keep their functional bosses happy. Yet, more than minor borrowing jeopardized the project. As a result, I had to keep a watchful eye over the people resources and often had to fight periodic battles with functional managers. The managers were not directly involved in the complex project, and, as a result, they had difficulty comprehending the impact of the "borrowing," particularly when the project deadline date was months away. I found the PERT chart an invaluable asset in fighting these battles. When a serious borrowing occurred despite my protests, I would recalculate the activity time associated with the particular job on which the borrowed person was working, do a computer run to determine how the project deadline would be affected, and then inform my bosses of the new date when the project would be completed. Because the project had to be completed by the assigned deadline or the Department would be forced to sign another one-year contract with the outside vendor, the altered deadline date usually got everyone's attention. The matrix-management issues and the difficulties of running a large project, caused me to search the Michigan State University library system for help, a move which, in turn, led to my first serious look at the management literature.

Being a female professional did have its problems. For one thing, sometimes managers in other departments were quick to blame me if there was a problem involving an interface between my project team and one of their male subordinates. For another, other project teams in the computer area consisted of all males, making it easier to engage in team-building activities, such as going golfing. My team also had the only female programmer. We managed to find a variety of fun things to do during coffee breaks and sometimes at lunch. Finally, although it was widely acknowledged that I was a superior performer, I was never moved to an official civil service management position. Reorganizations were frequent and during one 18-month period, I had nine different bosses. Many people throughout the Department often quietly mentioned to me that they thought it was "terrible" that I had not been promoted to a functional manager position. Fortunately, the bosses to whom I was directly assigned were for the most part very competent, respected me, and afforded me a generous amount of slack. We all understood that I was on the cutting edge of change and just a little ahead of my time.

Any feeling of dissatisfaction that I felt were somewhat mollified by the fact that I was promoted within the specialist track and was not discriminated against in pay. To the contrary, individuals with computer skills began to command a premium. I was moved to the computer specialist series, and given relatively high pay raises. One day reading a newspaper article, I was shocked to learn that my pay was in the upper one percent of women in the United States. Because I was only three years out of a master's degree program, the statistic actually said more about the dismal pay situation for women than it did about the level of my pay.

Whatever my troubles were, there were also many people who were ahead of their times in being supportive. During the three years that I worked for the State, I only met one other wife of a Michigan State student working in a professional job (there could have been others, but they were rare) in any State agency. I recognized that I had been fortunate to have obtained the job that I had and was grateful for the major responsibilities I was given. I learned a tremendous amount in a very short period. To this day, I am very grateful to the individuals who broke the norm in hiring me and the many wonderful individuals with whom I was privileged to work. They more than made up for the people who discriminated against me or tried.

The best outcome of the job was that I became fascinated with management problems. It was a challenge attempting to get diverse constituencies with vastly different perspectives (such as social workers and computer specialists) to work effectively together. My increasing exposure to the policy level also helped me realize how important it was to establish goals, let the workforce know what is expected, and put resources behind them. Some individuals seemed to be highly motivated to do a good job; others had to be cajoled. One of my more wild dreams was to someday advise Congress about the management feasibility of potential bills, so that bills could be fixed before being passed. In any event, there were many management issues of interest.

Bob and I had been talking for some time about the possibility of my obtaining a Ph.D. someday. But it would have been very difficult financially for both of us to have been in the early stages of Ph.D. programs at the same time. Also, I wasn't sure about the area in which I wanted to study. After he passed his comprehensive examinations, Bob also initiated a campaign to get me to apply to a Ph.D. program.

My growing fascination with management problems led me to consider the business school at Michigan State. Bob and I knew that women were not very prevalent in business schools and were concerned about whether or not I would be accepted. We were delighted when I received a letter of acceptance.

I still had the problem of being the project leader for the Medicaid Nursing Home project. I did not want to leave the many people who had helped me and who were on my team in the lurch. I had promised the project team that the project would be a success and my leaving before it was done would have jeopardized its chances. Fortunately, the Michigan State University Management Department gave me permission to start the spring semester on a part-time basis. My employer gave me permission to leave early one afternoon a week to take a graduate class. It was a major struggle running a project that required that I work a minimum of 50-60 hours per week and trying to take a doctoral course at the same time. But somehow I survived.

At the end of the spring semester, the project was implemented and was widely considered to be a success. One day shortly after implementation, the Budget Director for the State of Michigan got on an elevator on which I was riding. He greeted me and congratulated my team for being the only project team to date in a state agency that had successfully implemented a major computer system on time and within budget. This and other kudos that we received made the project team very happy.

I was scheduled to continue the doctoral program on a full-time basis starting in the fall. During July and August, my husband and I took a trip to what was then the Soviet Union. When we returned, there was a letter from Dalton McFarland, the chairperson of the Management Department, inquiring whether I would be able to accept an assistantship that had unexpectedly become available. Of course, I quickly agreed.

As I began the program as a full-time student, I was pleased to learn that another female doctoral student was just beginning the program (although she left the following semester to pursue other interests) and there were two women doing doctorates in marketing. My courses in organizational behavior and human resource management were as interesting as I had hoped, particularly those taught by Dalton McFarland, Henry Tosi, Winston Oberg, and Robert Penfield.

Management faculty members at Michigan State were generally supportive during my doctoral program. Richard Gonzalez, who had taken over as chair, continued Dalton McFarland's approach of treating me equally. Still the overall situation was somewhat difficult because, initially, I was somewhat ignored by many of the doctoral students. Fortunately, Henry Tosi would often come down to the coffee room in the College of Business to sit and talk informally with doctoral students. His willingness to accept me as part of the group ultimately made it difficult for others to exclude me. When I began to talk about preparing for comprehensive exams, many people appeared surprised. Apparently some thought I was just taking courses until my husband completed his program. After I took one of my minor exams and passed it, I was taken more seriously.

Doctoral students in the Management Department and a few in Marketing formed a "5:05 Club" which met, usually with spouses, at 5:05 every Friday night during regular semesters. We would gather to have pizza and the like at the Coral Gables, a local watering hole. Several of the most active members also had spouses that were in graduate school or had completed advanced degrees. Frequent attendance at the "Club" was both fun and useful in helping me integrate into the emergent informal group. In particular, Ramon Aldag, Jim Tartar, Hank Sims and their spouses made my husband and me feel welcome at the 5:05 Club.

During my doctoral program, Gwen Norrell, a Ph.D. who was head of the Michigan State University Career Testing Office, befriended me. I often enlisted her help in sorting out things that occurred because I was female from things that occurred for other reasons. Should I complain to the department chair and the dean that an associate dean from another business school came for a recruitment visit and interviewed every male doctoral student in our department who was on the job market, but did not interview me? (I did complain.) One day a ripped off magazine cover was taped to my office door showing a brassy-looking blond female cartoon figure saying to a forlorn-looking male, "Move over buster, I'm taking your job." Was it a form of harassment or was it an attempt to be funny? (Turns out the two well-meaning peers who put it up there thought that I would get a laugh out of it and were truly amazed to consider that it could possibly be interpreted as a hostile gesture.) During it all, I learned how difficult it can be when you are in a minority situation to separate out the way you are being treated because of your status (in this case, a doctoral student) from that attributable to being a minority member (a female). I am sure that I sometimes got upset when I shouldn't have and let things go by that I should have confronted, but I did my best to keep an even keel through it all and appreciate the efforts of many others to do the same.

I remember one funny incident in which Manton Gibbs, a male doctoral student with whom I shared an office one semester, had told his students that he would have his exams graded by a certain time. Unfortunately, he didn't show up at the office at the appointed time. As a result, for about an hour students had been knocking at the door and saying, "Is Mr. Gibbs in." When I said, "no," they would invariably say "Well, are you his secretary?" When Manny finally came in I said in an exasperated voice, "Manton, I wish you would tell *your* students that I am *not your* secretary." He looked at me a minute, grinned, and said, "I'll be glad to, Kay, if you will tell *your* students that I am not *your husband.*" I was astonished! I never occurred to me that *my* students were making that kind of assumption about *him.* Although I knew that people's assumptions about gender roles were causing me all kinds of problems, I hadn't fully comprehended the kinds of adjustments that other people also had to make. Many of the problems seemed to stem from people's prior expectations. I began to wonder how much of the fact that there were so few women in Management could be attributed to expectations that women would not perform well in such roles rather than to their actual performance. I thought it would be an interesting area of study.

While in the doctoral program, I conducted extensive research and wrote a paper on Soviet attempts to build a national network of computers to help with central planning and other administrative matters. The paper combined my expertise with computers with my knowledge of Russian studies and the Russian language. It was ultimately published in *Soviet Studies*, a prestigious international journal. During my doctoral program, I continued to have an interest in project management and its impact on individual behavior. Also

I enjoyed learning about various aspects of human resources management, which was one of my minor areas of study. I was particularly interested in issues at the intersection of organizational behavior and human resources management, such as equity theory applied to pay allocation decisions, expectancy theory applied to job choice, or attribution theory applied to performance appraisals.

During most of my doctoral program, I was unaware of how rare females in management doctoral programs really were. Thus, I did not initially fully recognize how far on the cutting edge of change the Michigan State University Management Department had moved in accepting me into their doctoral program. We would read journals, like the *Journal of Applied Psychology*, and often see female names as authors or coauthors. The realization of how few women scholars were in management came when I attended my first national Academy of Management meeting in 1971.

By that time, I was working on my dissertation, and was starting into the job market. I went to the registration desk to register and was handed one of those name tags with a plastic flap that fits inside the breast pocket of a man's suit coat. Because it was problematic for me, I asked if there were any pin-type holders that women could use. The woman who was helping me called over the person in charge of registration. He looked at me in shock and blurted out, "We never thought about any women coming." I returned his look of shock. I soon found that almost all of the women I had seen in the lobby were spouses of attendees. I only remember seeing maybe two or three women at most who seemed to have official name tags on, but I did not meet any of them. I do not remember any women being on the program in any capacity, but there may have been a very few.

For me and my spouse who was with me, the meeting was awkward. My husband vividly remembers walking into the job placement area with me. One of the people running placement rushed over to him and asked, "What kind of a job are you looking for? How can we help?" My husband pointed to me and said, "She is the one looking for the job in Management." The placement person looked amazed and somewhat askance.

During the meeting, the faculty from Michigan State, particularly Dalton McFarland (my doctoral dissertation chair) and Henry Tosi (a dissertation committee member) introduced me to people, including a number of well-known scholars. The scholars generally were extremely polite and very willing to talk to me about research. Job interviews, though, were rather tense, as many interviewers seemed unsure of how to handle the situation. The circumstances were made more difficult by the fact that my husband and I were a dual-career couple. Overall, people at the meeting were cordial if somewhat tentative—seemingly not quite sure whether my presence was a good or bad omen. I felt that I was given the benefit of the doubt, though, by a number of long-term prominent members of the Academy (such as Keith Davis and Harold Koontz) probably, in part, because my doctoral chair was Dalton McFarland, a highly-respected past-president of the Academy and past editor of the *Academy of Management Journal.* I continue to be extremely grateful that "Dalt" was willing to sign on as my doctoral chair at a point when it was not necessarily the popular thing to do.

Back at Michigan State, I was hard at work on a dissertation. After considerable thought, I finally decided that my dissertation would focus on the general question of whether males and females differed in their approach to leadership. An extensive library search turned up only a small number of relevant empirical studies with reasonably sound methodologies. Many of the available studies dealt with principals of elementary and secondary schools, settings which do not parallel the typical business situation closely enough to be useful. The assumptions in much of the popular press seemed to be that women were not cut out for management and that neither females nor males would be willing to work for female bosses.

A major obstacle to investigating possible differences in female and male leadership styles and outcomes was the relatively small number of female managers operating in business settings. After discussions with several professors and peer doctoral students, it appeared most feasible to conduct a field study using the groups of students playing a business game in the Management Department's introductory course in management. The setting did have the advantage of making it possible to appoint leaders and the game was already an integral part of the management course. Therefore, it was more of a field setting than a laboratory setting, yet manipulations were possible. Ray Aldag, who was teaching the large introductory course, kindly gave me permission to collect my data in its discussion sections, and took many steps to facilitate my data collection process.

The overall results of the study indicated that there were no significant differences between the leader behaviors of the female and male leaders—at least as perceived by their subordinates. There also were no significant differences in performance between female-led and male-led groups. In addition, satisfaction levels on several dimensions relevant to the task were generally similar for groups led by female and male leaders.

In completing the dissertation, Dalton McFarland, who as mentioned previously chaired the dissertation committee, gave me particularly useful guidance on conceptual approaches to the problem, and helped nurture my appreciation of theory. Henry Tosi, another of the committee members, spent considerable time working through the methodology with me and helping me learn how to develop research designs. Frederick Wickert, a faculty member from the Psychology Department and another committee member, aided me tremendously with the writing, teaching me how to carefully describe research methodology and results. Of ccurse, all of the committee members helped with all aspects, but each made particularly valuable contributions in certain areas.

As a result, the dissertation experience proved to be very excellent training and positioned me to be able to independently conduct research studies.

During the period that I was completing my dissertation, I was also going to other campuses on interviews. I received my first job offer from the University of Tennessee; but did not accept the position because it turned out to involve teaching mainly strategy, which was not really one of my major fields of expertise. One large stumbling block to my job search stemmed from the tendency of schools to focus extreme attention on where my husband was going to be employed. Some schools were reluctant to even consider me without knowing up front what he was going to do. Yet it was difficult for us to figure out where to focus our efforts when we could not assess even basic levels of interest. We eventually decided that I would take a position somewhere and he would work on turning his dissertation into a publishable book.

When I did go on interviews, there were sometimes questions that were difficult to handle. For example, the faculty member from a major state university picking me up from the airport posed the question, "Well, should we have our wives come to the dinner tonight?" (I sat there thinking, this is only the second interview I have ever been on, how am I supposed to know? I guess they had a similar problem, because I was probably the first female candidate they had ever interviewed.) Finally, I said, "Well, do they usually come?" He said, "no." I said, "Well, I would handle it as you usually do." (Subsequently I did not receive a job offer.) During an interview trip to a very prominent private business school, one of the faculty members confided to me after lunch that I was being interviewed only because the school was afraid of a law suit from the Equal Employment Opportunity Commission and that if I were hired all the faculty in the managment department would have to give up their pay raises for the year. Mercifully, I did not receive a job offer. However, they did later try to recruit me from the University of Maryland. I finally accepted a position at the University of Massachusetts, Amherst, which had a very good faculty group in management. Interestingly, the UMass management department was also able to hire the only other new female management Ph.D. who seem to be on the market that year, Mary Lou Bryant from Indiana University.

At the first business-college-wide faculty meeting of the fall semester at UMass, a faculty member proceeded to nominate me for the position of Faculty Secretary, the person who would record the minutes for the meetings of the approximately 90 faculty. Whereas I was certainly willing to do my share of service, I was unwilling to be thrust into such a gender-stereotyped role right in the door. In fact, I was quite upset, although I tried not to show it as all eyes in the auditorium turned toward me to see what I would do. As tactfully as I could, I declined the nomination, citing "obvious reasons." Next, though, the person who had nominated me was himself nominated and quickly elected. It was a show of support and I appreciated it.

At UMass, I quickly immersed myself in writing up the results of my dissertation for journal publication. Because the results of my dissertation suggested that male and female leaders may not behave much differently, I began to explore other avenues of explanation for the sparse number of women managers. One possibility was that females might not see business as instrumental to achieving desired outcomes. One study I conducted showed that female business majors saw business as significantly more instrumental to outcome attainment than did female psychology majors, reinforcing the potential usefulness of the expectancy model for analyzing female occupational choices and attitudes toward business.

Tony Butterfield, a colleague in the Management Department, and I began working on a couple of research projects. One of them attempted to look at another side of the gender-differences-in-leadership question. We wondered whether the same behavior by a male or female leader would possibly be evaluated differently by others. To test this prospect, we developed scenarios depicting different leadership styles and found that leaders engaging in "consideration behavior" were evaluated more favorably when the leader had a female name, than when the leader was depicted as male. On the other hand, "initiating structure" behavior was evaluated more favorably when the leader had a male name. Thus the very same behaviors were rated differently depending on the gender of the leader. Moreover, the leaders were evaluated more favorably when their leader behavior was congruent with gender stereotypes.

Meanwhile, Max Wortman, who was also a colleague in the Management Department, was able to gain access to a large hospital with a sizable proportion of female managers. Our study focused on the gender-differences in leadership question as part of a larger study that would provide information about the organization's training needs. Therefore, it was possible to ask individuals to provide information about various aspects of their leadership situation without calling direct attention to the gender issue. The results strongly supported the findings of my dissertation: few major differences in leadership behavior as perceived by subordinates. Interestingly, there also were few differences in satisfaction with various aspects of the work situation for individuals working with female versus male managers. This latter finding was particularly important in view of strong suggestions in the practitioner and popular literature that individuals working for females managers would experience lower job satisfaction than those working for males.

On the teaching side, some of the management faculty members at UMass, such as Joe Litterer, Halsey Jones, and Fred Finch, were experimenting with experiential learning, an approach to which I had had limited exposure. As I worked with them on various sections of MBA classes, they taught me a great deal about the art of teaching and the power of involving students in in-class exercises.

At UMass I came under intense pressure to be on various committees. On the positive side, there was a growing effort to make sure that women faculty were included in the governance of the University. On the negative side, like most universities, women faculty were in relatively short supply compared to the numbers of committees. I began to envision the prospect that my research time could be completely usurped by service requests and was determined not to let that happen. Even when I clearly was doing much more than my share of committee work, I still found myself in the very uncomfortable position of having to decline requests that I join still more committees. It was frustrating because many faculty and administrators were attempting to broaden opportunities for women faculty and I wanted to support their efforts as much as I could. Vice-Chancellor Robert A. Gluckstern was a prime mover behind attempts to involve more women in campus activities. Fortunately, he was sensitive to the need to protect research time for female faculty.

One day at the beginning of my second semester at UMass, a male student who was registered for one of my classes came into my office, sat down, looked me in the eye and said, "I just want you to know that there isn't anything about management that I could possibly learn from a woman." I wish I could report that I threw back a great retort, but actually I was *speechless*. Discriminatory behavior was not usually that blatant, particularly from students whom you would be grading, but it does serve to illustrate some of the problems with being on the cutting edge of change. (My Management colleagues and I amused ourselves for weeks thinking up things I could have or should have said.) Classes in business schools at the time typically contained a very high proportion of male students. On the other hand, the *Boston Globe* came up and did a story about my presence, because I was a relatively rare female professor teaching Management. Assistant professors do not usually generate that type of excitement, either.

When I first became a member of the UMass faculty, I officially joined the Academy of Management. Eventually I received a membership certificate with my name nicely printed in on the preprinted form. There was only one problem. The preprinted part of the certificate read in part "in recognition of *his* contribution." In other words, the membership certificates for the Academy essentially assumed that the members would be male. Apparently, a large batch of certificates had been printed at some point.

When I received the certificate, though, it was a frustration. Moreover, the certificate came on the heels of a letter from a journal editor (not *AMJ*) asking me to revise and resubmit my article on the subject of male and female leaders, but requesting me to please add a footnote stating that males had overseen the project so that readers could be assured that the study was not biased! These continuing gender issues were becoming exasperating!!

I was aware that a number of scholarly associations were taking steps to better the status of women. Given the situation demonstrated by the certificate,

I decided to write a letter to the president of the Academy of Management, who was then Charlie Summer of the University of Washington. In the letter, I complained about the certificates, asked that a committee on the status of women be formed, and offered to help. The letter had an attachment listing many other associations that had established such committees. I recognized that it was somewhat presumptuous for a new assistant professor to send a letter complaining to the president of the Academy of Management and that there were risks involved.

Within two or three weeks, I received a letter in the mail. I do not wish to reveal who wrote it, as no useful purpose will be served at this point. Suffice to say that it definitely was not Charlie Summer, who had shared my letter with a number of Academy people to determine what should be done. The letter, though, was written on stationary with the Academy of Management logo on it and seemed to be the official response. The one and one-half page letter contained a number of statements, such as:

> ... While you are new in the Academy and in your profession, I did feel that you hurled some implications in what seemed to be an angry tone. I'm sure that you try for a more objective view in your research than does the rather inflamatory, activist publication *MS*.

The letter dismissed my concerns stating in part that "hair-splitting over imagined insults is sub-professional and deplorable in my opinion." It also strongly implied that I was unwilling to work my "way up the ladder" and instead wanted "special" treatment. Although it was meant to be sarcastic, the letter contained the following prophetic statement:

> I hope that you will volunteer for some of the many openings we have in the Academy for workers of various sorts and if you put in a seven day week like most of the leaders of the Academy do, you may well become president of the organization some day....

The letter was a graphic illustration of the problem and I felt devastated when I received it. Although there was no indication in the letter, I learned later that a copy had been sent to each member of the Academy of Management Board of Governors. While I sadly pondered what to do next, the following day, I received a telephone call from Charlie Summer, telling me that a letter was on the way from him and telling me generally what was in it. He also told me that he had received a copy of the initial letter, had no knowledge that it was being sent, and wanted to apologize for it. Summer's letter, which arrived shortly thereafter, apologized for the certificate situation. He also indicated that the certificate would be redesigned, the idea of a Committee on the Status of Women would be placed on the Board agenda, and a questionnaire would be sent out soliciting interest from Academy members in serving on such a committee.

A few weeks later, Charlie Summer contacted me again and told me that he had met with Bette Ann Stead of the University of Houston during a regional Academy meeting and had enlisted her help. Working with Charlie, it was decided that Bette Ann and I would prepare a preliminary report on the status of women in the Academy. The report would be presented to the Academy Board of Governors meeting in August in conjunction with the national meeting to be held in Boston. Meanwhile, the mailing to the Academy produced several women volunteers for the committee, as well as the names of a few males. Subsequently, the Board appointed me as chair of the committee and, in addition, selected other committee members based on recommendations made by Bette Ann and I from the list of volunteers. The committee consisted of myself, Rosemary Pledger, Bette Anne Stead, Margaret V. Higginson, Alan Filley, and Don Hellriegel. (Alan Filley and Don Hellriegel deserve special thanks, because they were two prominent males who volunteered to be on the committee when the whole idea was not universally popular.) The committee was scheduled to meet in August at the national Academy meeting.

Meanwhile, Bette Ann and I worked on a preliminary report to present to the Board of Governors in August. Bette Ann was unfortunately unable to attend the Academy meeting that year, so I presented our report to the Board. The Board members were sitting at tables that formed a big square with space in the middle. Charlie Summer, George Steiner, and Rosemary Pledger were seated almost diagonally across from the place where I was asked to sit when I entered the room. After I sat down someone began to ask me a question and started out by saying "What shall we call you? Ms. or Mrs." To wit Max Wortman, a Board member and my colleague at UMass, piped up, "why don't you just call her 'Dr.'" Everyone laughed and that seemed to break the ice.

Based on the preliminary investigation, the report showed that no woman had ever held an elected national office or had ever been nominated for an elected position. Rosemary Pledger, though, had just been made administrative secretary, an appointed position. It appeared that no woman had ever served on the editorial board of the *Academy of Management Journal.* Of the 340 participants on the national program for the current meeting, only 5 were women; but among the 57 session chairmen and 58 discussants listed, none was a woman. Other similar data were included about the Division Chairs and Regional Division Officers. By looking at a membership list, we were able to identify 49 female members among the 2500 member Academy. It is possible that we missed a few who had first names that were ambiguous as to gender or who used initials.

As I presented the report, I remember Charlie Summer, George Steiner, Rosemary Pledger, Max Wortman, and Jack Miner reacting particularly positively. From where I sat, though, I could not see everyone well because some people were on my same side of the room and to my right and left. Rosemary continued to smile and shake her head at me reassuringly as I looked up from the report periodically to gauge the reaction around the room. I had never met Rosemary, but we were friends from that day on. Then, and subsequently, Rosemary was a strong champion for raising the status of women in the Academy. Overall, the Board was extremely gracious and constructive in receiving the report.

As the Academy meeting itself began and I would meet some of the spouses of Board members, I gained additional insight into the responses of the Board. Several of the Board members had shown their spouses the initial response letter that had been sent to me and the spouses were appalled! Thus, several spouses of Board members were instrumental in helping to initiate the effort to change the status of women in the Academy. Particular thanks is due to Carol Summer and Jean Steiner. Ironically, that initial response letter I received was the best thing that ever happened. It unwittingly established without a doubt that there was a real problem to be solved here.

During that same national Academy meeting, the newly formed Committee on the Status of Women was able to meet. The Committee decided to issue a similar but more extensive report that would focus on the coming academic year. The report would go into greater depth regarding Academy divisions and regions. Even in doing this report, we started to see evidence that the announcement of the Committee and the collection of information was beginning to have some effect. It turned out that the Southwest region had the most women members, with 19; whereas the Midwest had the fewest, with 9. The section on the divisions of the Academy showed that some divisions had no women members; none had women officers, but now though, two divisions had women on committees for the first time. More women began appearing on programs in slots like chairperson and discussant. Rosemary Pledger and I presented the new report to the Board of Governors the following August. It was accepted well. It appeared that women were invading the Academy with no noticeable detrimental effect. Bette Ann Stead took over the Committee the following year. Dorothy Harlow was another woman who became prominently involved in very early efforts to improve the status of women in the Academy. From that point on, the number of women participating in the Academy steadily increased, aided immeasurably by the many individuals who subsequently participated in the future Committees on the Status of Women and in the creation of the Women in Management Interest Group, which later became a Division. A few years later, Rosemary Pledger became the first female to be elected President of the Academy of Management.

While all this was going on, I tried to somehow stay focused on my research. The Division of Research at Michigan State University chose my dissertation for publication as a monograph. My husband and I did an extensive study of women in managerial and professional positions in the United States and the Soviet Union, which was published in *Industrial and Labor Relations Review*. Meanwhile my husband was called back to Michigan State to teach in the History Department. We spent the summer after my first year at UMass as Fellows of the Russian Studies Center at the University of Michigan in Ann Arbor.

My dissertation research and the paper that Max Wortman and I did comparing female and male leaders seemed to appear just as major questions were being raised in the popular press about whether or not men behave differently in leadership positions than women do. On one occasion, I was invited by Bette Ann Stead to be a speaker at a Women in Management Day at the University of Houston. There were several prominent women managers on the program. I spoke about recent research suggesting there may be few differences in the leader behaviors of male and female managers. At dinner, the women managers said they had always thought of women managers as more sensitive and caring than male managers, but after hearing my talk about some of the emerging research findings, they were revising their views. They said they could all think of prominent examples where the opposite was true. The discussion ended with the notion that maybe it was best not to make assumptions based on gender about how managers were likely to behave. In view of their collective experience and the emerging research, it seemed to be good advice at the time. In the ensuing years, I have yet to see research findings that are sufficient compelling to contradict this stance.

I should emphasize that in pursing possible gender differences in leadership behavior by managers, my interest was purely scientific. From a personal point of view it really didn't matter to me whether there were or were not differences. In any event, the emergence of contingency theories of leadership was pointing to the need for different types of leaders.

One problem at UMass at the time was resources. Amherst turned out to be a fairly expensive place to live on an assistant professor's salary and I usually had to use personal funds to pay for even basic expenses related to research, such as long-distance phone calls, photocopying and the like. Although I liked my colleagues in Management very much, I found it fairly difficult to line up research sites or interact with businesses because of our distance from major cities like Boston.

Part way through my second year at UMass, I was approached by a friend from my doctoral program at Michigan State, Rod Chesser, who wanted me to come to Syracuse University. After a couple of visits, Syracuse made me a good offer with a promise of more research support. Although in many ways, I hated to leave UMass, it looked like a good opportunity. The following fall I joined the Syracuse University faculty as an associate professor. When I moved to Syracuse, my husband was still in Michigan. Whereas the reaction of colleagues in the field seemed to be one of astonishment, the move did have one tremendously helpful side effect. From that point on, schools would approach me about my interest in particular jobs without focusing on my husband.

Syracuse University seemed to operate with many fewer committees than had been the case at UMass. Thus, there was a more workable service load. Syracuse University had an unusually good rapport with the people who lived in the region. Faculty were afforded an amazingly high level of respect in various interactions, ranging from giving speeches to picking up one's dry cleaning from a local establishment.

At Syracuse, I wrote several papers with Rod Chesser on organizational climate, an area in which he had a particular interest. I also began some work on job orientation, investigating the values that individuals place on various job rewards—again searching for reasons for the low representation of women in business and managerial positions. Although there were some significant differences between male and female business majors in job orientation variables, it appeared they might be less than between business majors and individuals in other fields—in this case psychology. This study built on some previous work by Phil Manhardt at Prudential Insurance Company. After contacting me about my work, Phil and I began to collaborate on follow-up research that considered the job orientation of a large pool of college graduates beginning work in a major organization over a nine-year period. We found that females gave significantly less emphasis to career objectives and significantly more emphasis to work environment and interpersonal job aspects than did males. Interestingly, an analysis of trends over the nine-year period showed a convergence of female preferences toward those of males on the two dimensions on which gender differences had been found. During this same period, two doctoral students, Chuck Evans, who is now a Professor at Florida A&M University, and Mel Stith, who is now Dean of the College of Business at Florida State University, and I wrote a review of the literature comparing black and white leaders. The paper was published in the *Academy of Management Review*. It highlighted the relative dearth of relevant studies and pointed to a number of methodological difficulties that needed to be overcome.

At Syracuse, I also began to do research involving computer professionals, an interest related to my earlier work as a computer project leader. Some of my work focused on professionalism and how it relates to such outcomes as organizational commitment, job satisfaction and turnover among computer professionals. One of the interesting findings was that the computer specialists were more satisfied with various aspects of their jobs, more committed to their organizations, and less likely to leave when the organizational reward system gave significant weight to various aspects of professional behavior (e.g., ability to work without much guidance, high concern for client or user interests, and participation in professional organizations). Congruence between the reward system and professional norms seemed to be important.

Based on the gender-related research in which I was directly involved and emerging findings by others, it continued to appear that women did not behave much differently than men when they were appointed to leadership positions in business-related situations. Rather, the problems seemed to be more in the area of women not obtaining leadership positions. When my review article dealing with the sex structuring of organizations appeared in the *Academy of Management Review*, it gained a great deal of attention. In one part, the paper reviewed literature on gender differences related to leadership style, job satisfaction of leaders and subordinates, and job performance. The review concluded that differences attributable to gender in the areas reviewed generally tended to be nonexistent or too small to account for the sparse representation of women in managerial ranks. The paper also used Edgar Schein's career stages as a framework within which to identify possible causes for the gender structuring of organizations and identified a number of questions in need of further research. Recently I was honored to receive the Sage Scholarship Award for significant scholarly contributions to gender-related research from the Women in Management Division of the Academy of Management. The *Review* article was mentioned in particular as having had considerable impact on gender-related research.

One of the continuing problems with resolving questions of either gender or race issues in leadership is the fact that developing a definitive concept of leadership has been illusive. In pursuing the gender question further, I found myself becoming more and more heavily involved in questions of what is leadership? As I began to consider interesting leadership questions for research, I had to keep reminding myself that I had not set out to resolve the whole issue of the definition of leadership. Rather, I was interested in the issues of why there weren't more women in leadership positions and how to gain greater access for women to leadership positions.

Because my husband had meanwhile moved to the Washington, DC, area, I began spending summers there and enjoyed the area very much. Management faculty at the University of Maryland kindly provided me with access to the library and computer facilities, so I visited the campus periodically. In one discussion with faculty, Carl Anderson, Craig Schneier, and I decided it would be interesting to explore possible sex differences in motivation to manage as a potential explanation for the relatively small proportion of female managers. We began two research projects in that area. Our research, completed later, indicated that although males scored significantly higher than females on motivation to manage, a rather small amount of the variance was explained by gender.

The strong research focus at Maryland was especially appealing to me. When the possibility of an opening in the Management and Organization Department at Maryland arose, I was asked if I was interested in being considered. Ultimately, Steve Carroll played a major role in bringing me to Maryland. Marty Gannon and Ed Locke were also heavily involved and I continue to appreciate them as colleagues. I had enjoyed my three years at Syracuse University and my colleagues in Management, including Rod Chesser, Don

DeSalvia, and Gary Gemmill. It was a productive time for me and it was with considerable regret that I left.

Faculty members in the Management and Organization Department at Maryland traditionally have been strong researchers, making it a particularly good atmosphere in which to work. Moreover, I was very fortunate that Robert L. Gluckstern, who had been Vice-Chancellor at UMass when I was there, had become Chancellor at the University of Maryland. He was leading a strong effort to involve women in various campus activities and was encouraging the hiring of women faculty both within the College of Business and Management and elsewhere on campus. Over the next several years, in the course of our hirings, the Management and Organization Faculty was able to add two new excellent female faculty members, Judy Olian and Susan Taylor.

My interests in the gender question had also led to my involvement in the Social Issues Division of the Academy of Management and a broader paper on corporate social performance and policy with Ray Aldag, who by now was at the University of Wisconsin. At the invitation of William Frederick, University of Pittsburgh, I became head of the Division's Research Committee and was eventually elected Chair of the Division. True to the values of its domain, the Division was particularly open to having women participate in the governance of the Division. Although my current research has taken a somewhat different turn, I remain a member of the Division and am grateful to the many members who welcomed me and allowed me to contribute. Gradually, I also performed other roles in the Academy, such as serving as a member of the Constitution and By-Laws Committee and was eventually elected to the Board of Governors.

Over a period of time, I served on a number of journal review boards, including those of the *Academy of Management Review* and the *Journal of Applied Psychology*. My work on the review board of the *Journal of Vocational Behavior* coupled with the motivation to manage research and my earlier interest in job orientation led me to consider the vocational choice literature as a possible source for theoretical frameworks that could be used to explain the relatively low proportion of women in management.

At around this same time, David Martin, who had recently retired as a Brigader General from the U.S. Army and is now a Professor in the business school at American University, joined our doctoral program. After several discussions about management theory, Dave and I discovered that we shared many common perspectives and research interests and began to collaborate on a number of research projects, including a couple of studies in the vocational choice area. One of our studies continued earlier work on motivation to manage, but focused on MBA students, finding that males scored significantly higher than females on motivation to manage, with much of the difference attributable to the low scores of part-time female MBA students. Overall, however, the scores of female and male MBA students were rather high, suggesting that motivation to manage may not be a major issue inhibiting the upward mobility of female MBA students. In another study, we evaluated the usefulness of Holland's Vocational Preference Inventory (VPI) and the Myers-Briggs Type Indicator (MBTI) in predicting MBA areas of study concentration. The VPI proved to be a potentially useful tool for career counseling. We also found, though, that the relative numbers of females and males choosing the different concentrations (e.g., accounting, marketing, finance.) were fairly close to their proportions in the MBA program. There did not seem to be major gender differences in this area.

Based on an invitation from Richard D. Ashmore and Frances K. Del Boca to write an assessment of gender differences in leadership for their book on the social psychology of female-male relations, Dave Martin and I completed another major review of the literature. Overall, we found that the literature still seemed to be pointing to relatively few differences between female and male leaders, at least in business-related settings. Even when differences were found, they were, for the most part, relatively small.

Given the continuing findings indicating few differences, my own interests began shifting to other issues. For quite a period I was heavily immersed in gender issues in organizations from a personal standpoint because of being at the cutting edge of changes in the role of women in management. At the same time, I was also heavily involved in gender issues from a research point of view. Both because of the dual immersion and because answers to some questions were beginning to emerge, I began to feel a need to take a different perspective and work on some other management issues for awhile. It is important to note that I never set out to spend my entire career studying gender issues, though they continue to interest me. Instead, my concerns have always been somewhat broader—more in the realm of how to make organizations places where individuals, regardless of their gender or other background factors, could maximize their own development and also help keep their organizations on the cutting edge of change. Many interesting research questions fall within this broader concern and many of them have linkages to both human resource management and organizational behavior. Moreover, many of the problems underlying the underrepresentation of women in managerial positions seemed to be related less to leadership differences than to issues of selection, performance appraisal, and fair compensation. Many of these issues fall within the purview of human resources management.

Dave Martin and I found that we particularly shared common interests in linkages between organizational behavior and human resources management. As a result, we began to collaborate on research in the areas of performance appraisal and compensation. Because of Dave's concern with legal issues and performance appraisal, we conducted two reviews attempting to understand the role that performance appraisals were playing in discrimination-related court cases. The reviews suggested problem areas in performance appraisal that, coupled with applicable theory, have formed the basis for some of our on-going work.

In the compensation area, we developed a theoretical paper applying the resource dependency approach to help explain managerial pay allocation decisions. The major premise was that managerial pay allocation decisions do not stem simply from a manager's assessments of performance, but are significantly influenced by the degree to which the manager is dependent on particular subordinates and by threats to that dependence. Two follow-up studies we have conducted so far support the resource dependency model. In one study involving pay allocations made by bank managers, for example, our results indicated demands for higher pay from a boss were ineffective unless a subordinate was in a situation in which the boss was highly dependent on the subordinate (in this case, the subordinate had expertise in an area that was critical to the bank's strategic plan). In another study, involving managers in the industrial uniform industry, results again supported the dependency approach with managers awarding higher raises to an employee on whom they were dependent for expertise. Moreover, political connections led to a higher pay raise only when a subordinate made a dependency threat (expressed dissatisfaction and threatened to take the matter to higher levels). These results are interesting because they helped to document that managers are influenced by factors other than performance in allocating pay. The findings also are useful in advancing theory building related to compensation allocations.

After I was at Maryland for three years, I was promoted to Full Professor. The following year, I agreed to become Chair of the Department. Later that year, I learned that I had been elected Vice President and Program Chair of the Academy of Management. Traditionally, the Vice President and Program Chair automatically becomes President Elect the following year, and then moves to President of the Academy of Management. Thus, that initial letter I received, when I filed a complaint about sexist Academy membership certificates, ironically turned out to be prophetic in stating that ..."if you put in a seven day week like most of the leaders of the Academy do, you may well become president of the organization some day ...."

Being Program Chair was exciting and time-consuming. It was a chance to influence the character of at least some of the national program, and also an opportunity to work closely with a number of Academy Division Program chairs, who were usually excellent scholars. Among other things, we were able to initiate the first All-Academy Symposia in an attempt to provide further integration to a program that was rapidly becoming differentiated into more narrow divisional interests. We also continued to stress joint symposia involving two or more divisions or interest groups. Collecting all of the materials, producing the program, and preparing for and running the annual meeting, though, was a major logistical challenge. When annual meetings run smoothly, the job of the Vice President and Program Chair appears to be simple. I certainly gained a new appreciation for all the unseen efforts of past and future national Program Chairs. Because the meeting was scheduled for Dallas, I received a great deal of help with logistics from Claire Cunningham of the Cox School of Business at Southern Methodist, who agreed to serve as local arrangements chairperson through contacts kindly made by John Slocum.

The following year, I served as President Elect and spent a great deal of time on Academy of Management business. Unfortunately, I also received the sad news that my mother had cancer. Because of her illness and my Academy responsibilities, I found it necessary to step down as Chair of the Management and Organization Department. Gratefully, colleagues in the Management and Organization Department, particularly Steve Carroll and Marty Gannon, sometimes filled in for me at classes so that I could periodically go to Michigan to visit with my mother. Unfortunately, medical science was unable to stem my mother's cancer and it was a major loss when she passed away three months after I became Academy President.

One of the major issues that emerged during my Academy presidency was whether or not to move forward with establishing a new journal, the *Academy of Management Executive*. The idea for the journal, which had emerged during the previous year when John Slocum was President, was largely supported by the Board, but had generated some questions from a few Board and Academy members. The purpose of the publication was to aid in transferring technology in the sense that it would contain articles highlighting the practical implications of recent advances in management theory and research. It could also be a conduit for obtaining articles and opinion pieces from executives. The idea had merit and I wanted to support it. In my Presidential Address, titled "The Best Kept Secret," which I gave at the national Academy of Management meeting that year in San Diego, I argued that the Academy had much to offer, but that the Academy and its members needed to be more proactive in making research findings known to executives. Moreover, we also needed to do more listening to the perspectives executives have to offer to maximize advances in knowledge about management. Ultimately, the Board decided to go ahead with the *Executive* and when a membership vote was necessary to put the *Executive* Editor on the Board of Governors more than 90% of the returned ballots approved the measure.

Holding each of the offices of Program Chair, President Elect, and President was extremely time consuming, but also rewarding in that I was able to influence future directions of the field in a number of ways and also was able to interact with the many fine members of the Academy. During the years from Program Chair to President, many people in various Academy positions were exceptionally helpful to me. Among them were Walt Newsom, Secretary-Treasurer; Donna Ledgerwood, Advertising Director; and Carolyn Dexter, Director of Membership.

Shortly after completing my term as Academy President, an opportunity arose to write a management book that would solidly build on the latest management research, but make it accessible to students through careful explanations and ample linkages to business examples. The book was to be published by McGraw-Hill. After considerable thought, Dave Martin and I decided to undertake the project. It proved more formidable and time-consuming than we had anticipated, taking several years to complete. The project did have the advantages of affording us an opportunity to review a wide spectrum of literature and a chance to closely read the literature relating to creativity and innovation, one of the themes in the book. When students come back after a summer job, an internship, or taking a job after graduation and talk about how valuable they found what they have learned, it all seems very worthwhile.

Along the way, I have worked to become the best teacher possible within my capabilities. I gradually began to attend meetings of the Organizational Behavior Teaching Society and for a period served on its Board of Directors. Some of the Society's workshops were particularly valuable in helping me recognize more clearly the difference between content and process in the classroom and beyond. I also have learned a great deal from my husband, who is a master teacher. I have come to recognize that teaching to the best of one's ability is a life-long journey and a challenge of continual improvement. I devote a considerable amount of time to my teaching obligations.

One of my long-term teaching interests has been instructing a management course for undergraduates. Part of our role as scholars in my view is to disseminate our findings in ways that will aid others. Helping undergraduates, as well as graduate students, develop is part of that obligation. We fortunately have many very fine undergraduate students at the University of Maryland and it has been my privilege to devote at least some of my teaching load to interacting with them.

Over the course of time, I have been honored with College of Business and Management Krowe Awards for teaching excellence and have been named a University of Maryland Scholar-Teacher, a university-level award for excellence in both teaching and research. Most importantly, however, my students are learning more and show some enthusiasm for a field that I find fascinating. I have especially enjoyed interacting with a number of doctoral students over the years, including Ming-Jer Chen (Columbia University), Cris Giannantonio (Chapman), Kay Tracy (Gettysburg), Karyll Shaw (Villanova), Jim Guthrie (Kansas), Amy Chesney (Georgia Tech), Bill Fitzpatrick (Villanova), Cynthia Lee (Northeastern), Marilyn Gist (Washington), Shannon Davis (North Carolina State), Mike Ostrow (Howard), Kline Harrison (Wake Forest), Judy Scully (Florida), and, more recently, Julie Kromkowski, Suzanne Masterson, Amy Kristof, Amit Gupta, Cathy Durham, June Poon, Nags Ramamoorthy, and Craig Pearce. Julie Kromkowski and I have a research study in the works. Monica Renard (West Virginia) and I are just completing a project dealing with pay-equity issues and part-time work. At the present time, I serve with enthusiasm on the Maryland Business School's Teaching Enhancement Committee, which has as its mission orchestrating activities that will aid our faculty in developing their teaching talents to the fullest extent possible.

Sometimes challenges arise that you do not anticipate. One year ago, my left leg had a full-length cast on it because of a broken ankle and a knee that had just been through surgery. A velcro cast was on my left wrist, which was also broken. Someone carelessly left a puddle of soapy water on the floor of the main foyer of our new business school; no warnings were posted. I was the unfortunate person who happened along. It has been a year in which I experienced great pain, particularly with the extensive physical therapy I had to undergo to get my knee to bend. I am still working on some of the problems, but I am grateful to be walking. My colleagues in the Management Department were helpful to me during this crisis, especially Judy Olian and Susan Taylor who took over my classes despite their own busy schedules.

Still, within a week or so after my return from the hospital, my husband had moved a computer and a fax machine to the downstairs study, where I had to stay for almost three months because I could not climb stairs. Thus modern technology and his resourcefulness enabled me to continue working, at least to some extent. Steve Carroll helped locate a talented undergraduate, Susan Sheets, who did a tremendous job helping me several hours per week with my work, including retrieving things from my home office, which was upstairs and inaccessible to me. I returned to teaching on crutches during the Spring semester and continue to be grateful to my family and the many colleagues and friends who came to my aid and helped me overcome the challenges that I faced.

As I am ending this autobiography, it almost seems premature. According to the longevity charts, I likely have a considerable number of years left in which to make contributions. It is hoped that someday a sequel to this autobiography will be able to report that I made substantial progress on continuing work at the cutting edge of major issues in Management. I would also want it to say that I persisted in developing my capabilities, was successful in aiding others in developing theirs, and used my talents to try to bring about change for the better. At the same time, I would want it to recount that I devoted considerable time and attention to helping my husband, the rest of my family, and the colleagues and friends who rallied to my side when I needed them. Not surprisingly, my advice to readers would be to do your best to develop your own capabilities, stay around supportive people who encourage you to grow, and try to use your capabilities to make a positive difference—it is very motivating!

PUBLICATIONS

1972

Whither Soviet computer centers: Network or tangle? *Soviet Studies, 23*, 608-618.

1973

Male and female leaders of small groups. East Lansing: Division of Research, Michigan State University.

With R.A. Bartol. Soviet information-handling problems: The possibilities in computer usage. *Computers and Automation*, June, 16-18.

1974

Male versus female leaders: The effect of leader need for dominance on follower satisfaction. *Academy of Management Journal, 17*, 225-233.

1975

The effects of male versus female leaders on follower satisfaction and performance. *Journal of Business Research, 3*, 33-42.

With R.A. Bartol. Women in managerial and professional positions: The United States and the Soviet Union. *Industrial and Labor Relations Review, 28*, 524-534.

With M.S. Wortman, Jr. Male versus female leaders: Effects on perceived leader behavior and satisfaction in a hospital. *Personnel Psychology, 28*, 533-547.

1976

With M.S. Wortman, Jr. Sex effects in leader behavior self-descriptions and job satisfaction. *Journal of Psychology, 94*, 177-183.

Expectancy theory as a predictor of female occupational choice and attitude toward business. *Academy of Management Journal, 19*, 669-675.

With D.A. Butterfield. Sex effects in evaluating leaders. *Journal of Applied Psychology, 61*, 446-454.

Relationship of sex and training area to job orientation. *Journal of Applied Psychology, 61*, 368-370.

1977

Factors related to EDP personnel commitment to the organization. *Computer Personnel, 7*, 2-5.

1978

With D.A. Butterfield. Evaluators of leader behavior: A missing element in leadership theory. In J.G. Hunt & L.L. Larson (Ed.), *Leadership: The cutting edge*. Carbondale: Southern Illinois University Press.

With C.J. Evans & M.T. Stith. Black versus white leaders: A comparative review of the literature. *Academy of Management Review, 3*, 293-304.

With R.J. Aldag. An appraisal of recent empirical studies in corporate social performance and policy. In L.E. Preston (Ed.), *Research in corporate social performance and policy* (Vol. 1). Greenwich, CT: JAI Press.

The sex structure of organizations: A search for possible causes. *Academy of Management Review, 3*, 805-815.

1979

With M.S. Wortman, Jr. Sex of leader and subordinate role stress: A field study. *Sex Roles, 5*, 513-518.

Individual versus organizational predictors of job satisfaction and turnover among professionals. *Journal of Vocational Behavior, 15*, 55-67.

With P.J. Manhardt. Sex differences in job outcome preferences: Trends among newly-hired college graduates. *Journal of Applied Psychology, 64*, 477-482.

Professionalism as a predictor of organizational commitment, role stress, and turnover: A multidimensional approach. *Academy of Management Journal, 22*, 815-821.

1980

Female managers and quality of working life: The impact of sex-role stereotypes. *Journal of Occupational Behaviour, 1*, 205-221.

With C.E. Schneier. Sex effects in emergent leadership. *Journal of Applied Psychology, 65*, 341-345.

With C. Anderson & C.E. Schneier. Motivation to manage among college business students: A reassessment. *Journal of Vocational Behavior, 17*, 22-32.

An addendum to the sex structuring of organizations: The special case of traditional female professions. *Journal of Library Administration, 1*, 89-94.

1981

With C. Anderson & C.E. Schneier. Sex and ethnic effects on motivation to manage among college business students. *Journal of Applied Psychology, 66*, 40-44.

Vocational behavior and career development, 1980: A review. *Journal of Vocational Behavior, 19*, 123-162.

1982

Manuscript characteristics as viewed by editorial review board members: Lethal and non-lethal errors. In D. Loeffler (Ed.), *Understanding the manuscript review process: Increasing the participation of women.* Washington, DC: American Psychological Association.

With D.C. Martin, Managing information systems personnel: A review of the literature and managerial implications. *MIS Quarterly*, Special Issue, 49-70.

1983

Turnover among DP personnel: A causal analysis. *Communications of the ACM, 26*, 807-811.

1985

With C.E. Schneier & C.R. Anderson. Internal and external validity issues with motivation to manage research. *Journal of Vocational Behavior, 26*, 299-305.

With D.C. Martin. Predictors of job status among trained economically disadvantaged persons. *Psychological Reports, 57*, 719-734.

With D.C. Martin. Managing turnover strategically for positive results. *Personnel Administrator, 30*(11), 63-73.

The best kept secret. *Academy of Management Newletter, 15*(4), 7-19.

1986

With D.C. Martin. Women and men in task groups. In R.D. Ashmore and F.K. Del Boca (Eds.), *The social psychology of female-male relations* (pp. 259-310). New York: Academic Press.

With D.C. Martin & A. Lyons. Human resource management: A growing necessity for DP managers. *Journal of Systems Management, 37*, 32-36.

With D.C. Martin. Expectancy theory as a predictor of turnover among the economically disadvantaged. *Journal of Social and Behavioral Sciences, 32*, 13-25.

With D.C. Martin. Training the raters: A key to effective performance appraisal. *Public Personnel Management, 15*, 101-110.

With D.C. Martin. Holland's VPI and the Myers-Briggs Type Indicator as predictors of vocational choice among MBAs. *Journal of Vocational Behavior, 29*, 51-65.

With D.C. Martin & M.J. Levine. The legal ramifications of performance appraisal. *Employee Relations Law Journal, 12*, 370-396.
Making compensation pay. *Computers in Personnel, 1*(2), 35-40.

1987

With D. Koehl & D.C. Martin. Quantitative versus qualitative information utilization among college business students. *Educational and Psychological Research, 7*, 61-73.
With D.C. Martin. Managerial motivation among MBA students: A longitudinal assessment. *Journal of Occupational Psychology, 60*, 1-12.
With D.C. Martin. Potential libel and slander issues involving discharged employees. *Employee Relations Law Journal, 13*, 43-60.

1988

With D.C. Martin. Influences on managerial pay allocations: A dependency perspective. *Personnel Psychology, 41*, 361-378.

1989

With D.C. Martin. Effects of dependence, dependency threats, and pay secrecy on managerial pay allocations. *Journal of Applied Psychology, 74*, 105-113.

1990

With D.C. Martin. When politics pays: Factors influencing managerial compensations. *Personnel Psychology, 43*, 599-614.

1991

With D.C. Martin. *Management*. New York: McGraw-Hill.
With D.C. Martin. The legal ramifications of performance appraisal: An update. *Employee Relations Law Journal, 17*, 257-286.

1992

Pay systems as strategic mechanisms for promoting continuity and innovation. In S. Srivastva (Ed.), *Executive and organizational continuity*. San Francisco: Jossey-Bass.
With L.L. Hagmann. Designing team-based pay plans: A key to effective teamwork. *Compensation & Benefits Review*, November-December, 24-29.

1994

With D.C. Martin. *Management* (2nd ed.). New York: McGraw-Hill.

1995

With D.C. Martin, M. Tein, & G. Matthews. *Management: A pacific rim focus.* New York: McGraw-Hill.

Janice M. Beyer

Performing, Achieving, and Belonging

JANICE M. BEYER

My first memory illustrates the deep ambivalence I felt, even at four years of age, over the expectations others placed on me for exceptional performance. I was at my paternal grandmother's house for her annual birthday party. This was an especially exciting occasion for me each year because, although we visited my grandmother fairly frequently, we saw relatively little of the cousins, aunts, and uncles on that side of the family. What I remember about this particular year's party was being asked to sing in front of everyone. A much older cousin's fiancee—a dark curly-haired Irish policeman I thought very handsome—had been talking to me in the awkward but kindly way that adults talk to children when my cousin suggested to him that I should be asked to sing "Deep Purple." He looked at me intently and asked. I wasn't afraid or angry at the request. I knew the song, I was apparently accustomed to such requests, and I wanted to please this attractive and nice man. But for some reason I did not want to sing. My parents were not in the room, and so, perhaps for the first time, I decided to resist. Because I was normally an obedient child I faced a quandary. I didn't feel I could just refuse. It didn't occur to me to simply say no. Nor did it occur to me to cry. I was too proud for that. Suddenly I got a bright idea—I would *pretend* to sing. I can still remember how satisfied I felt when I thought of this solution. I stood in the middle of the room, as requested, and mouthed the words of the song but did not utter a sound. The adults kept saying, "Sing louder. We can't hear you." I didn't comply. Instead I continued mouthing words until I had finished the song.

Management Laureates, Volume 4, pages 39-84.

ISBN: 1-55938-730-0

In its way, this incident represented a real triumph to me and that is undoubtedly why I remember it. I had asserted my own preferences. I had resisted successfully—but not directly. It would take me a long time to learn that I could simply say no. And even longer to see some requests as presumptuous or inappropriate.

Fortunately, my cousin did not hold my refusal against me—she probably attributed my silence to my being shy. She subsequently invited me to serve as a flower girl at her wedding to the handsome policeman about a year later. The pictures of the event show a bashful little girl wearing a silly pancake-like hat with ruffles all around who didn't smile because of missing front teeth.

As I reflect upon it now, I realize that a pattern had obviously been set early in my life that, since I learned easily, I would be singled out, taught to perform, and then expected to perform. While I have certainly enjoyed the feelings of mastery and self-efficacy that being able to perform in various arenas has brought to me, I have often felt ambivalent both about being singled out and about what others expected. My parents showed me off because they were proud of my accomplishments. But their pride and approval had mixed consequences. I was glad to be able to please them in these small ways. I was compliant and internalized their desires for me to achieve. What I didn't learn about was what I liked and wanted. Sometimes I knew what I didn't want and then I could resist in some acceptably round-about way. But I generally complied because I wanted the approval it brought. To the degree I knew my own desires I tended to compromise them to what seemed the stronger and more articulated desires of others that I cared about. I grew up eager to please others but knowing little about how to please myself.

The result has been a patchwork of accomplishments that reflected the expectations of others in my life as much or more than it reflected my own desires or inclinations. Looking back, I feel I have had several quite different lives—perhaps because I have taken on a series of identities determined much more by my significant others and my circumstances than by own aspirations or intentions. In another account of my career, I have described that process as "Going with the Flow" (Beyer, forthcoming).

GROWING UP: MY FAMILY AND NEIGHBORHOOD

Clearly, by the time I found a way to avoid singing "Deep Purple" for an audience, a pattern of expectations had already been established in my family and among my relations: Janice performs difficult feats; Janice is exceptional. I don't remember learning the song, but I must have been coached. I doubt I simply picked it up from listening to the radio. The family stories of my early life I heard as I grew up were always about my performances of some sort or another. Janice could recite the Gettysburg Address when she was two years old. Janice was toilet trained before she was a year old. Janice walked at 9 months and talked full sentences before she was a year old.

Janice also had to stand up on a chair during meals to gain attention—mine was a family that liked to talk and, in the process, shared a lot during meals. In particular, all of the family members ate dinner together and discussed what had happened to us that day. My parents spent more time talking than we children did, and as I sat through those meals, I undoubtedly absorbed their values toward work and their work organizations. In these conversations, my father was usually complaining a bit about his bosses or talking about what went awry; by contrast, my mother was always doing more than expected, being successful, and being praised.

My family was German on both sides and more working class than middle class in its attitudes and origins. My father, who had only an eighth-grade education, was a foreman in a box factory during my childhood and later started his own small steel rule die business. My mother, who had completed only a two-year secretarial course in high school, stayed home when my sister and brother were small, began working part-time when I was three and full-time after I entered first grade. Her initial job was as a cash register clerk in an A&P supermarket, but she fairly soon rose to be assistant business manager of that store, then business manager, and eventually—during World War II—she became the only woman store manager in Milwaukee.

My mother's working was very unusual in our neighborhood, and the other mothers regularly teased me about it. Her parents had emigrated to the United States from Hungary when she was three years old. Unfortunately, although he tried a variety of different jobs, my grandfather was never a financial success. He was, however, an intelligent and well-informed man. He became active in the Progressive party, and eventually found his calling as a politician. He ran for state senator and lost, but later ran for city alderman and won. He served as alderman of the area in which I and all of my relatives lived during most of my childhood and until shortly before his death, surviving consolidations of the common council and redistricting with a reputation for complete honesty and service to his constituents.

My grandmother had eight children—only six of whom survived childhood. She relied heavily on my mother, who was the oldest girl, for help. Probably because she was so bright and capable, my grandmother not only burdened my mother with much of the care of her younger siblings, but also with many other family responsibilities. Her siblings responded with great respect and love for my mother, treating her from the time I can remember more like a mother than a sister. In June of 1925, when my grandmother was expecting her last child, my mother and father married. He was 21 and had been working full-time since he was 11. My mother was only 16, but had finished a two-year secretarial course in high school and been working full-time as a secretary in a lawyer's office for about a year. Both were accustomed to the responsibilities of adulthood. One set of photographs of their courtship that I remember seeing as a child show a very pretty young woman with a beautiful smile and a very dark-haired skinny young man dressed in big bulky sweaters and wool pants standing in front of a snowy hill with ice-skates in their hands.

Soon after they were married, my father, with the assistance of some of the relatives, built a very small two-room house on the edge of the city where land was cheap. This house, surrounded by huge poplars that dwarfed it, was remodeled to provide four rooms before my birth and was the family home of my childhood. In their early years there, my mother carried water from a neighbor's well, tended a large garden, and raised chickens so she could sell the eggs. She and my father struggled to make ends meet and hold on to that house during the Depression, when my father's work hours were cut. But he never lost his job entirely and my parents were the ones who lent money to my grandparents and other relatives. My parents could always be counted on to do whatever needed to be done, and to do it well. My father had mechanical abilities and was a hard worker. My mother could cook an excellent meal, bake, sew, or rise to whatever life seemed to require. She sat up with those who were ill or dying, and then made all of the funeral arrangements. She was always there for members of the family. She was so generous with her efforts that I internalized her selflessness in a way that probably did not serve me in good stead. For most of my life, I lived the philosophy that it was selfish, perhaps immoral, to deny a loved one their heart's desire. So I gave whatever seemed to be required to help them reach their dreams. That might have been a fine philosophy if others had shared it. As a child I didn't see that my mother was also very good at getting back in one form or another what she had given. I internalized the giving part but missed seeing the getting part.

Given these personal qualities and her general intelligence, my mother ended up being the driving force in her own family and also in ours. She was, in many ways, a remarkable role model for the times. I am sure that if she had not dared to work and been as successful as she was at both work and family, I would probably not have had the drive to pursue a career at a time when most women didn't. My mother's working, however, was not motivated by desires for a career. She started working during the Depression to provide financial security and a better standard of living to the family. Later, when I was in high school and the family was better off, she stopped working. She was then about forty years old—not much older than I was when I received my Ph.D. degree and started my career. My father had started his own business by then, and she did his bookkeeping, but she never really worked full-time again.

Somewhere during her childhood my mother had also acquired middle class aspirations and ideas. My father's family was very narrow-minded, but my father greatly admired my mother and was happy to go along with her ideas. She saw to it that we went regularly to museums, the zoo, and community events like concerts, the circus, the state fair, the Fourth of July and other holiday parades and celebrations. We made several car trips to see national parks out West. Later, after we children were grown, my parents traveled extensively throughout the world. My mother bought books as gifts for my older siblings and thought she was building a home library. In truth we had only 6-8 books in the library—classics like *Oliver Twist, Kidnapped, Little Women, Little Men*, and *Treasure Island*—all of which I read many times over. Our library wasn't much, but it was more than many of my playmates had. My parents also read the newspaper regularly and kept well-informed about current events. Since I was such a great reader, I was soon reading the newspaper, too. But mostly I read books from the public library, taking them out five at a time, the maximum allowed. In these and other ways, I grew up relatively rich in cultural advantages compared to the other working-class children I knew.

My sister, brother, and I were all given music lessons. My brother was talked into trying the violin. My sister and I played the piano. I was the only one who persisted with even minimal practicing and was willing to continue them for more than a year or two. In general, I was the relatively compliant child in the family; my older siblings rebelled and resisted my parents' influence, especially after they reached their teens. Because I was much younger—seven years younger than my sister Betty Jane, five years younger than my brother Elmer Warren—I could observe the results of their rebellions from a very different developmental stage, and early realized that they hurt themselves as much or more than they hurt my parents. I didn't understand why they were rebelling but determined that I would behave differently and did.

Of course my mother wanted all of us to attend college. My father did not disapprove, but was less convinced of the value of higher education. As it turned out, my sister rejected the idea altogether and took an office job instead. She was only an average student and had been encouraged to take secretarial courses in high school. My brother wanted to be a lawyer and probably could have been a very good one. He entered a pre-law program at a local Catholic university but pursued his studies half-heartedly. He was living at home and torn between his desire to go out in the evenings with his neighborhood friends, who were working full-time and not in school, and doing his schoolwork. He didn't have late afternoons available for studying because, as expected by my parents, he was also working part-time. He ended up spending too little time studying, wasn't getting good grades, and eventually dropped out of college, perhaps because he realized he would soon be drafted anyway because of the Korean War. He was. He served in an Army accounting unit in Korea. When he came home from the war he took a few night courses in accounting but never went back to prepare for law school or earn a college degree. Instead he went to work for my father in his steel rule die business. I watched all of this happen and resolved I would not let anything similar happen to me.

GRAMMAR SCHOOL

Of course, school was another arena in which I could achieve and be exceptional. In my school pictures, I look like a very shy child. But I must have stood out early, for I was the one who was selected to have the only speaking part for a kindergartner in the Christmas play. When the time of the play arrived I had the measles and never got to take part. I only completed half of first grade and was then skipped to second grade. The skipping meant I started second grade in January, and when my parents decided to transfer me to Catholic school, I had to be either kept back or skipped another half grade. The nuns agreed with my mother than I should be skipped, which meant that I began third grade when I was seven. The rest of my school life I was a year younger than my classmates, and always somewhat of an exception for that reason alone. I was smaller and skinnier than the other girls until I was in high school. So I heard taunts of "Beyer, Beyer, skinny as a wire" from the boys.

In general, however, I managed to keep a pretty low profile during grammar school. I didn't try to stand out—I wanted to fit in. It was difficult because I lived so far away from school and no one else from my neighborhood attended the same church school. One benefit of my mother's working was that I was left pretty much to my own devices, as long as I got home by supper time. My sister, who was the oldest and therefore supposed to supervise me to some degree, was not eager to do so, and my parents apparently trusted my ability to take care of myself. This freedom allowed me to sometimes walk home with schoolmates to their neighborhoods so I could spend some time playing with them. I made my way home later in the afternoon by a combination of buses, streetcars, and walking. I learned a great deal of self-reliance early.

During the later years of grammar school my parents transferred me from a neighborhood piano teacher to one at the Wisconsin College of Music. My new teacher, Irma Habeck-Dufenhorst, was very ambitious for her students and competitive with the other piano teachers in the city. Soon after I began taking lessons with her, I found myself challenged to play in numerous formal recitals, at piano contests, at the local park band shell, and even at the state fair. I usually felt vaguely inadequate playing at these public events. I didn't feel I had outstanding musical abilities and my hands were too small to play the bravura romantic-period pieces that audiences most appreciated. But I had some finger dexterity and a good memory. And I became accustomed to performing. In eighth grade, the month I turned 13, I was pushed into giving a whole recital on my own. I played a Mozart piano concerto and about ten other pieces from memory.

Because my piano teacher exaggerated my accomplishments, my parents began to see my future as a piano teacher. Their intentions were the best—they were envisioning a future that would enable me to combine a career with marriage and children. They thought I would be able to teach piano students in my home and take care of my children at the same time. At the time, it seemed like a good idea to me, too.

THE HIGH SCHOOL YEARS

When I began high school, in 1947, my mother laid down a new set of expectations. She drove me to school the first day, stopped the car in back of the school, and explained that there was a club called X Club at the high school and that, in order to belong to this club, students had to have all grades over 90 on their report cards. She also told me that she expected me to belong to X Club. When I got a grade of 87 in algebra on my first report card she went to speak to the teacher. I never got a grade under 90 again, and at graduation, I was valedictorian of my class. The truth was that academic subjects were easy for me—much easier than music.

Largely because of my parents' financial and emotional investments in my musical training and their expectations that I would end up teaching music, I majored in music in high school. I was required to take either orchestra or band and sing in one of the school choruses every semester. I chose orchestra and learned to play the violin in a rudimentary way. I sang in a succession of choruses, each year qualifying for the next most advanced one. As a music major, I was also required to take dance classes as part of the physical education requirement. I thoroughly enjoyed both singing and dancing, perhaps because both were group activities, and both came easier than playing the violin. Also, my high school held several performances during the year in which I could participate with my classmates. We had lots of fun rehearsing and performing together.

Soon after I began high school my English teacher approached me to ask me if I would like to sell advertising for the school paper. I was flattered to be asked and agreed. It was a pretty nitty-gritty job that involved walking throughout the business streets of the section of town near my high school and stopping at numerous barber shops, hardware stores, and other small businesses to induce their owners to buy small ads for our bi-weekly paper. I not only sold the ads, but was obliged to take them the issue of the paper with their ad in it to collect their payment. The ads cost only 75 cents per inch, and many advertisers took only an inch or two per issue. To collect these small sums I walked the streets in that section of town at least every two weeks for years. When I go back to Milwaukee now, and drive down those same streets, I still remember many of the businesses I used to visit so regularly. Some of them are still in business in the same location.

My efforts selling advertising earned me the title of Assistant Business Manager of the newspaper and yearbook during my junior year, and in the first semester of my senior year, I was named Business Manager of the newspaper. Much more important to me than the titles I earned was that these activities gave me a place in the select group of students who comprised the editorial board of the paper. Not only did we have the privilege of taking all of our study halls in the student newspaper office, but every two weeks, we were excused from classes to spend a whole day at the printers proofing the paper. There must have been some vacancy on the business staff because I got to go along to the printers beginning in my freshman year. All of us who got to go considered this a great plum, not only because we were excused from classes, but because we had free time between sets of proofs that we could spend playing cards and joking with each other. I remember tasting Chinese food for the first time on one of those occasions. What satisfaction it gave me to belong to and be accepted by this insider group of upperclassmen. We considered ourselves very sophisticated and intellectual compared to the other groups of student leaders—the student government and athletic types.

I found the business side of the newspaper less interesting and exciting than the editorial side, and as soon as I was a junior and could take newswriting courses, I began writing for the paper, too. I was soon busy as a reporter as well as an advertising salesperson. These efforts culminated in my being named Editor of the paper in the second semester of my senior year.

Working on the high school paper, I now realize, attracted me because it provided a possible way to have my efforts and achievements assist me in gaining entry to a group I admired. I felt more comfortable devoting my efforts to group goals than to standing out as an individual. Perhaps then any excelling I did would be accepted and appreciated rather than resented. By this time I had occasionally experienced what appeared to be envy and resentment from some of my peers. As it turned out, my efforts on the paper did give me a sense of belonging and brought me valued friendships, but some resentments and envy occasionally surfaced from time to time on the newspaper staff, too.

My musical activities in high school also put me into settings where I could achieve for and with a group. Because I was not an especially good musician and did not have an especially good voice, I was in no danger of excelling to an exceptional degree. I could simply be part of the group. In both orchestra and chorus I thoroughly enjoyed being part of a group effort. Singing in the a cappella choir, the most advanced choir in the school, was especially rewarding because we got to perform in the yearly operetta and some of us were selected to sing in various all-city choruses. I was not an outstanding singer, but a good enough musician to be selected for these activities. I never had an individual singing role in the operettas, but was good enough to be included one year in a small vocal ensemble number. Most important for me was that I was part of the group as we traveled downtown on the streetcar for a rehearsal, ate together, or just fooled around. Performing the music was inspirational, too. I still remember standing as the snow fell around us singing the Hallelujah Chorus for a sunrise Easter service at a local park.

My piano playing also took on more social aspects when my teacher encouraged another student, Colleen Kell, and I to perform two-piano works. I enjoyed those rehearsals and performances much more than I had enjoyed my solo performing. Less enjoyable were the piano lessons that my parents induced me to give to small children during my senior year in high school. They put a sign on the house:

Janice Mary Beyer
Teacher of Piano

for all the world (and my friends) to see. I didn't have the nerve to tell them that I didn't want to teach piano. But the experience of doing so hardened that resolve, and I decided I must do something else with my music education—it was unthinkable that I would simply abandon it. Too much time and money had been spent.

A solution presented itself when a reporter from one of the local newspapers came to a career counseling session at the high school and told us that those of us who wanted a career in journalism should not take journalism as a major in college, but instead major in a substantive area in which we would later be expert. He mentioned political science and economics. But I was thinking music. Now I knew what I could do—I could become a music critic on a newspaper. The leap wasn't hard to make since my only college-educated relative—an aunt—was art critic for the same local paper.

This solution provided me with several handy justifications for avoiding things I didn't want to do. I argued to my parents that I should not take music education and that ruled out attending colleges in Milwaukee and justified going to the University of Wisconsin at Madison because only the School of Music there offered a history and theory major. I knew from my brother's experience that it would be difficult to attend college while living at home. I would also lose many of the benefits I saw as connected with going to college—the social life and the freedom from parental scrutiny and supervision. That was to be the biggest benefit of all—to begin to live away from my family and to make more decisions for myself.

It helped to persuade them when I was given a gold medal at graduation, selected to be commencement speaker, and awarded several small scholarships that I could use at any school I wished to attend. My commencement speech was all about the gratitude that I and my fellow-students should feel toward our parents and the sacrifices they had made while raising us.

Later, in the 1950s, when I was in college and had gained some perspective on these events and my first glimmerings of the idea that women were not supposed to stand out as achievers, I cringed when I reflected on how well

I had satisfied my mother's expectations while in high school and came to resent those expectations. I also cringed at that time to think of what I had chosen to say during the commencement address. How dutiful I had been! Later I came to realize that the pattern of achievement those expectations has instilled in me opened many doors that otherwise would have been closed to me.

Looking back, I now realize that my mother's ambitions and the behaviors she modeled were a mixed legacy. Without them I would probably not have aspired to the career I have had, have gained the opportunities to pursue it, or developed the stamina to see me through the rough patches along the way. With them I have stood out as different from other women of my generation and borne the costs that went along with standing out and being different.

THE COLLEGE YEARS

I had enjoyed being part of the newspaper staff so much in high school that I joined the *Daily Cardinal* staff at UW as soon as I could. By the end of my freshman year I was one of the daily News Editors, Co-Magazine Editor for the weekend magazine page, and the music critic. I was probably spending 40 or more hours weekly in the *Cardinal* offices. My academic subjects were simply a background activity. Again, I felt accepted as a valued, contributing member of the group. Again I earned special privileges from belonging—this time that I did not have to keep the curfews that other freshman women did. It went without saying that, like my high school newspaper staff and real journalists everywhere, we thought we were very intellectually sophisticated and politically aware. We probably were in some respects.

In my sophomore year, interpersonal conflicts—triggered by the activities of two ambitious men students who wanted to control everyone else—rocked the *Cardinal* staff. I and many other members resigned. Working at the *Cardinal* wasn't fun anymore. The two troublemakers struck me at the time as simply trying to take power away from the chosen Editor. But perhaps they were more in touch with what journalism is about than the rest of us. They were the only ones who went on to become successful—indeed rather famous—journalists.

Meanwhile, I had become active in a new cause on campus—integrating the arts—and was more than busy organizing the first inter-arts festival for the following spring. No one knew exactly what integrating the arts meant, including me, so the job involved dealing with lots of disagreements about fuzzy ideas and motivating people who did not agree perfectly with what we were doing to help with it anyway. The festival was held at the end of my sophomore year, was a reasonable success, but didn't change anything about the arts on campus permanently. In the process of bringing it off, I had to raise contributions, appear on local television talk shows, arrange for the insurance and transport of paintings, and generally accomplish a plethora of other practical tasks. I also composed and conducted background music for the translation of a Chinese play that was performed at the Union. In the process of doing all of these things I learned a lot about how the university was run and even managed to win a political battle over the objections of the full-time director of the Student Union. The festival we had planned had apparently intruded into what he considered his turf.

THE LASTING BENEFITS OF MY MUSICAL EDUCATION

Although the relevance of music to scientific inquiry may not be obvious, I think I learned a lot in my musical studies that transferred to my scientific research, professional activities, and teaching. My early experiences playing the piano in front of crowds taught me a great deal about how to handle nervousness and other emotions in public settings where the actual costs to me were not high. I did not consider myself a great musician and so any mistakes I made were not too damaging to my ego and became learning experiences. Also, I was only a child and did not always fully appreciate the stakes involved. I took the idea of performing to these crowds as just one more thing to do because I was expected to do it. The idea that I could not never crossed my mind. I ventured and, if I was not a great success, I did not fail. I came out of these experiences with a certain assurance.

My interest in music history and theory taught me to be interpretive and analytical. The study of music history gave me an appreciation for the causal connections among social context, personal biography, people's ideas, and the form they take. The study of musical theory taught me how to analyze particulars and draw a general pattern from them. It also taught me to think analytically. I was good in these subjects and received lots of encouragement from my professors. They made me feel that I might have more than average analytical abilities.

Because I was a music major, I also had to take required courses in subjects in which I did not and seemingly could not excel. I had to arrange pieces for concert band, learn to play the cornet and drums, compose music and then listen to it performed in public, and direct a student orchestra. I was better at some of these tasks than others, but not confident about my ability in any of them. It was evident soon after I began music school that I had no inborn exceptional musical talent as some of my fellow students did, and so I had to learn self-discipline, persistence, and a degree of humility to successfully complete these courses. I probably should have switched majors, but I would have lost a whole year of credits if I did, and I was afraid of what my parents' reaction would be. So I stuck the course and managed to graduate with honors, but my record would probably have been far better if I had majored in almost anything else.

In a conventional academic sense, my undergraduate years were largely wasted. I did not study many subjects that I wish I knew more about now. I earned a Bachelor of Music degree, which required 98 out of 130 credits in music. I was required to take English, history, and languages, and I took my electives in skill-developing courses related to my plans to be a journalist. There weren't many credits left over to devote to traditional liberal arts subjects. I managed to sneak in some extras over one summer and by being exempted from a music course—a course in astronomy and two in English literature. Mine was far from a liberal education. What I did learn in college was how to handle myself in many different situations and settings and how to accomplish what I set out to do. I also broadened my intellectual horizons in general ways and, thanks to the people I came to know at the *Cardinal* and in the arts, I was exposed to new value systems and social perspectives. The values and perspectives I was exposed to at Wisconsin were liberal ones—much more liberal than that of my parents—and I hold them to this day.

COMBINING SCHOOL AND MARRIAGE

The summer after the inter-arts festival my whole world changed. I fell in love with the man who would shape my early adulthood. Tom Lodahl was the roommate of my best friend in Music School, Duane McGough. Duane and I studied together a lot and I had heard about various of Tom's annoying traits and peccadilloes. Our first date was during the final exam period of my sophomore year. Whatever we had each learned about one another from Duane became incidental during that first date. We were powerfully attracted to each other and fell deeply in love. Within a few weeks we knew we would marry within a short time. We didn't even have to discuss it or make a formal decision. We just knew. That fall my parents announced the engagement and Tom and I decided to marry the following year—at the beginning of our senior year. We did, on September 4, 1954, in a big Catholic church wedding, complete with the long white dress and veil and a big reception.

The timing of our marriage was a notable departure from Tom's original plans. He had not planned to marry until he had finished a Ph.D. degree in psychology. But after meeting me, he changed his mind. Although several years older, he was a semester behind me in school as a result of his military service during the Korean War. His career plans had jelled while he was in the Army working in a personnel placement office. He resolved to take extra courses so we could graduate together but did not sacrifice his preparation for graduate school. He was well-informed about what he should take, and carried advanced German courses and the math courses his small-town high school did not offer. By his senior year he was well-qualified to take advanced courses in psychology and statistics.

My career ambitions did not change in any essential way during our engagement, but I decided to expand my options by taking courses in radio and television writing and production, as well as newspaper writing. As part of these courses I wrote the Saturday afternoon newscasts for the local NBC affiliate and ran television cameras when the PBS station in Madison first went on the air. In my courses in television production, we wrote, directed, and produced many short programs as class exercises. I also took a course in critical writing and remember panning Marilyn Monroe's performance in the film thriller *Niagara* in one of my written assignments. My arguments apparently did not impress the professor, for she only gave me a B on that review. I suspect my writing revealed more moralistic disapproval than anything else, for I did not approve of the role of sex object even then. When I saw the film replayed recently on television I reflected again on what a naive girl I must have been. But I also saw why I had been repelled as a young woman by the sexy image that Marilyn projected for men to enjoy.

In my junior year, in accordance with Tom's and my marriage plans, I took a part-time job so I could save money for our school and living expenses after we married. In those days it was unthinkable that my parents would continue to pay them. Because a girl I knew in high school was married to a graduate student who worked there, both Tom and I had the opportunity to go to work in the lab of the famous psychologist Harry Harlow. Tom went to work for that graduate student; his job consisted mainly of running tests on monkey's cognitive abilities. My work at the lab consisted largely of calculating correlation coefficients and other statistical tests on an electronic calculator. I didn't know any statistics. I just did the calculations as I was instructed to do. But I felt the excitement of doing science around me, and enjoyed hearing Tom and the graduate students talking about the research.

Late that spring a shipment of pregnant rhesus monkeys that Harlow had ordered arrived. Among the graduate and other students who worked at the lab, the accepted explanation for his buying the pregnant females was to get two monkeys for the "price" of one. Because they were considered sacred, the Indian government would only allow a certain number of rhesus monkeys to be shipped out of the country in a given year, and they were thus in short supply. The newly arrived monkeys were given physical examinations before they were allowed to join the colony, and tuberculosis was discovered. The decision was made to keep the pregnant females in isolation so they could not infect the rest of the colony and use euthanasia on each immediately after she had given birth so that she would not infect her offspring. Tom was one of the students who administered the overdoses of anesthetic to the mother monkeys. He hated the job but felt it was necessary. One of the women graduate students was given the job of caring for the newly born monkeys, who had to be fed around the clock. She consequently named the first born Millstone, and followed that with Lodestone, Rhinestone, etc.

The lab was crowded, so the infant monkeys were kept in cages on a large table in a room that all of us passed through to get from one part of the lab to another. It was impossible to go by without stopping to watch them. They were amazingly like human infants in many ways and very endearing with their huge, intelligent eyes. That summer, shortly after they were born, the whole lab moved to a new and bigger building. Soon thereafter, Harlow's famous mothering experiments began. In them he systematically removed baby monkeys from their mothers to study their subsequent behaviors. I have always suspected that these experiments were not planned before the tubercular mother monkeys were put to death but grew out of Harlow's observations of their suddenly orphaned offspring. If I am right, it was an important instance of serendipity in science. I don't know for sure because by the time the mothering experiments began, I had a different job and had no opportunity to discuss the issue with his graduate students. No one, certainly not me, would have dared to ask Harlow directly.

Tom worked at two jobs the summer before we were married and became very skinny and tired. But our big wedding went off as scheduled and we honeymooned in Madison for a week fixing up our apartment before school began. Because I didn't like working for my friend's husband, I decided to find a different job. I soon found one assisting a bookkeeper at a small private airport. It was a small operation with just a grass field, but it was the local Cessna distributor and the owner was highly respected. While working there I had the opportunity to observe his leadership and the family-like culture it created. He was beloved by his employees and respected as a highly knowledgeable, forthright, and ethical man. He acted like the father of a huge family, even lending his employees money for down payments on their houses. I never saw him angry or short-tempered. Everyone but me had worked there a long time. In short, he epitomized the paternalistic employer in the best sense. I think I have carried that image of leadership with me as an ideal ever since. As it turned out, the bookkeeping I learned there would also stand me in good stead in the future.

Moving toward Tom's interests, I moved away from mine—journalism, music, and the arts. I began to be attracted to science, about which I had known very little before. I took my first and only psychology course after I met him. I served as a pretest subject in one of his experiments. I found the subject matter fascinating. I had always been a keen observer of people and now I began to see there were systematic ways to study and understand their behavior.

RECONCILING PARENTHOOD WITH TOM'S CAREER PLANS

A few months after Tom and I married I became pregnant. Once we realized this, Tom and I felt we had to reexamine our plans. We had planned to go to graduate school at the same time—he in psychology and me in musicology. I had given up the idea of being a mere journalist and took on aspirations similar to his. In truth, the idea of being a college professor was not entirely new to me, but I would probably never have dared to pursue it if I had not met and married Tom, who saw such an ambition as achievable and knew what to do to achieve it. To us, the pregnancy meant that we could not both go to graduate school at the same time. Tom's GI Bill entitlement was exhausted. It was before the days of student loans. As mentioned earlier, it never occurred to us or to my parents that they could continue to support me. Therefore, one of us would have to work. Tom went out for job interviews and was counseled to complete the Ph.D. instead. We saw Tom as the eventual breadwinner and, as such, felt he should not have to give up his plans to be a college professor. The only way that would be possible was if I worked after the baby was born. I acquiesced to this decision so readily, perhaps even instigated it, because I felt Tom would be bitter and unhappy if he gave up his ambitions. I did not reflect very long about how I would feel in the long run about giving up mine.

So Tom applied to graduate schools, and with Harlow's recommendation behind him, was accepted everywhere he applied. We sat down one evening, made a chart by which to weigh the decision, developed a list of criteria, and rated each school on each criterion. In this super-rational way we chose the psychology department at the University of California in Berkeley. Because I was expecting in August, we drove out to Berkeley to get settled right after graduation in June, 1955. Tom got a job welding in a truck body factory, and I cooked and napped and waited for the baby to be born.

While I was pretty apprehensive before she arrived, after Claire was born I was sold on motherhood. She was surely the most beautiful and remarkable baby that had ever been born. Tom, who was normally not very demonstrative, was also surprisingly carried away. We had already decided what must be done. I would begin looking for jobs as soon as I felt able, and Tom would plan to stay home with the baby as much as he could. He had wangled a teaching assistantship because he felt that would give him more free time during the day than a research assistantship would have. When Claire was a month old I began looking for work, and when she was five weeks old, I found a job—bookkeeping—at a radio and television station in San Francisco. Soon thereafter we found a great woman to take care of Claire at her home while Tom was in classes.

We followed this pattern throughout Tom's three years at Berkeley. I held several different jobs, none of them related to my college education. He took care of Claire when he could and when he couldn't, took her to our babysitter's house where she learned to be with other children. Tom applied himself very assiduously to his studies and had a remarkable memory. He was able to pass his French, German, and the three preliminary examinations in record time.

In his second year he also managed to co-author several papers that were accepted for publication. That summer he collected data at the United Airlines base at the San Francisco airport for a project with Lyman Porter and prepared to take his comprehensive exams in the fall. His progress was probably faster than the faculty had expected was possible and it may not have been in Tom's best interests. But he did it because it seemed the right thing to do at the time. He passed the comprehensive exams that winter because the faculty committee failed him on his first attempt. In-between the two sittings of the exam he came down with pneumonia. Despite these obstacles, he managed to design and collect data for his dissertation by the following August. In just three years from his Bachelor's degree, he earned a Ph.D. from the Berkeley psychology department and took a job as an assistant professor at the Sloan School at the Massachusetts Institute of Technology.

The years at Berkeley had a profound impact on me. We socialized a lot with the other graduate students and through hearing their conversations, I began to long to pursue the behavioral sciences myself. I picked up a lot of the values the students were being taught as if by osmosis. Also, Tom talked a lot about what he was doing and shared his concerns and ideas with me. I was beginning to feel very unfulfilled and was eager to take my turn in graduate school when he graduated. I decided I wanted to study sociology—the societal level it encompassed seemed to fit with the way I thought and the kind of perspective I had learned in music history. But when I told Tom about my intentions, he was disapproving. He told me that I was trying to compete with him. I did not think I was, but I didn't want any trouble in my marriage and decided to go ahead and pursue musicology instead, as I had once planned.

BEING A YOUNG FACULTY WIFE

Going back into musicology was a poor decision for me. I applied to the music school at Brandeis University, which had an excellent faculty, was admitted without difficulty and even given a fellowship. My fellow students were friendly and supportive. But I again felt inadequate and untalented and that I did not belong. My professors were again approving of my analytical efforts and abilities, but their approval only made me feel worse. I felt that I was getting by with my general intelligence and that I didn't belong in music. Also, with a husband and child to care for, I found myself less interested in music than I had been when I was unencumbered and undistracted by the pressures of everyday life. Everyone else in music school seemed to be spending every waking minute on something connected with music. I was not. Meanwhile, Tom began to ask if I wanted another child or not. Of course I did. I thought there was plenty of time. I was only 25. But Claire was now four years old, and people had been asking me for at least two years about why I didn't have another child. In my third semester of part-time study I knowingly allowed myself to became pregnant with our second child. By the beginning of the next semester we had bought out first house and were moving in. I was feeling tired and hassled, so I applied for a leave of absence, wanting to keep the door open but knowing that I would probably never go back. That June Andrea was born and I took on the role of a full-time wife and mother.

The years while Tom was an assistant professor at MIT were hard ones for me. I was disappointed and frustrated by my attempt at graduate school. I found the people in the Boston area relatively cold and unfriendly compared to California or Wisconsin. I felt I was losing my identity as a person. I was Tom's wife and Claire's mother, but what was I as a person? When we went out socially the men I met only talked to me about the weather, or asked me about how many children I had and what I thought of the schools where I lived. The women I met mostly talked to each other about matters related to running a household and managing children. It was the 1950s with a vengeance. One party I attended with some older faculty couples from Harvard and MIT was especially distressing. I already felt ill-at-ease and out-of-place as I listened to conversations about trips to Europe, skiing, and other things outside my experience. When I finally felt constrained to speak and remarked to the Dean's wife sitting next to me something about the hair style we shared (hair pulled back in a bun), she responded with impatience, saying, "What's the matter with you young women today. Can't you think of anything to talk about but children and other boring domestic concerns?" The answer, which exploded in my brain but was never uttered, was, "No, I can't." And I began thinking about why I had come to this pass.

Soon thereafter I became active in the League of Women Voters in the town of Arlington, where we lived. My experiences in the League were very gratifying—I was immediately put to work investigating and speaking about proposed local legislation, and my efforts were appreciated. Again I was working toward group goals and being appreciated and admired for what I did. I remember one of the women I worked with commenting to me that she and the other women had never seen a couple so much in love as Tom and I. Another sign that we must have made a favorable impression occurred when we were asked to usher for MIT's Centennial Ball. Only two couples were chosen from each school for this honor, and I felt reassured that I must have personally made the grade socially for us to have been selected. I never doubted that Tom would make the grade.

Soon thereafter, Tom had several nibbles from other schools interested in hiring him. He felt unsure of his future at MIT, and we didn't especially like living in the Boston area. He found it as interpersonally cold as I did. So he considered the possibilities and eventually decided to join the faculty at Cornell. In the summer of 1962 we moved to Ithaca, New York. There I would finally have the opportunity to pursue graduate work in the social sciences.

THE YEARS IN ITHACA

The years at Cornell had profound effects on all of us. Tom went as an untenured associate professor, earned tenure the next year, and soon thereafter was promoted to full professor. I earned my Master's and Ph.D. degrees. The girls grew up there, became proficient horsewomen with walls full of ribbons and trophies, and both earned undergraduate degrees at Cornell. Claire went on to earn a doctorate in veterinary medicine and married there, as well. In Ithaca we had our happiest times as a family.

Tom began his appointment at Cornell with great enthusiasm. He originated the first Ph.D. program in organizational behavior in what was then the Graduate School of Business and Public Administration. He designed and taught new courses. He served for two terms as editor of the *Administrative Science Quarterly*. He was instrumental in hiring both Bill Starbuck and Karl Weick to the Cornell faculty so that he could hand the editorship to each of them in turn. In these and many other ways, he was a unselfish and contributing citizen of the Cornell academic community.

When we moved to Ithaca, Andrea was two and Claire was seven. The next year Andrea began nursery school in the child development department at Cornell and I took a job editing in electrical engineering. I was able to turn that job into a part-time one the following year, and continued with it until I again began graduate school. The professors I worked for all had big grants, and I observed a lot about academic entrepreneurship and the influence it can bring while working there. I remember best one of the young full professors who had about half a million dollars in grant money yearly and thus was accorded a lot of deference by administrators and jealousy by his peers. He told me regretfully once that there were so many things he had wanted to do when he was more junior that he had not yet been asked to do, and now that he was asked, he didn't have time to do them all. This same man could not understand why I, a faculty wife, would want to attend graduate school. It was the mid-1960s and Cornell still had a nepotism rule.

From the time we arrived in Ithaca, Tom and I had a very busy social life. Patricia Cain Smith, who remembered Tom's application to Cornell for graduate school, held a party in his honor soon after we arrived. At that event we met most of the social scientists on the campus. We took full advantage of the opportunity, for Tom liked meeting and entertaining people, and so did I. Most weekend nights we were entertaining someone at our home or being entertained at someone else's. Ithaca had no memorable restaurants and few night spots. Few wives worked. So people entertained each other in their homes. We also got together as families, taking our children to the local parks, swimming, or to athletic events. We became especially close friends with many of the other psychologists on campus. We also got to know many of the arty types because one of our neighbors was librarian for the art and architecture departments. Later, after we had moved to our second house in Ithaca so our girls could have their horses at home, we joined the Yacht Club.

But long before that, I was liberated. Soon after we moved to Ithaca, I read Betty Friedan's book *The Feminine Mystique*. At the time, the ideas it contained were revolutionary. Until I read it, I had no idea that other women felt as frustrated as I did. None of the women I knew ever talked about such ideas. We had all been socialized in the 1940s and 1950s to feel that we should be happy as wives and mothers and not want anything else. A few, with children in school, held some kind of "assisting" jobs; they were research associates paid out of soft money from grants while their husbands were the professors. The exception was Pat Smith, and she did not have children. After reading Friedan's book, my resolve to attend graduate school in the social sciences became stronger. I didn't really envision a career—I simply wanted to use my mind. Tom, ever one to be at the forefront of ideas, seemed to embrace the new feminist ideas. Now he could accept my desire to attend graduate school in sociology. I decided to apply for admission to the rural sociology department at Cornell in the fall that Andrea would begin first grade. I chose rural sociology because I was pretty sure that, with my Bachelor of Music degree as undergraduate preparation, I would never be admitted to the Cornell sociology department in the College of Arts and Sciences. As it turned out, I didn't get admitted to rural sociology either because of the small number of new students they could admit. They expected I would wait for a vacancy at midyear, but I did not want to wait, and applied belatedly to the Organizational Behavior Department in School of Industrial and Labor Relations. Organizational behavior was what I really wanted to study all along, but I had felt constrained from applying to the ILR School because all of the faculty were our close friends. I applied for the Master of Science program, rather than the Ph.D. degree, because I envisioned myself, like the other faculty wives, assisting others in carrying out their research. This kind of career appealed to me because I thought I could become part of a group effort of the kind I had witnessed in Harlow's lab and in the electrical engineering school at Cornell. Also, a voice somewhere inside told me it was safer for my marriage.

BECOMING A SCIENTIST

I was admitted to the ILR School as a nondegree candidate and told I would have to prove myself by taking a full load of courses and earning all A's. I began the program in September of 1966, satisfied these requirements, and was admitted as a regular student and given a research assistantship the following semester. I asked to work with Gerry Gordon, a new faculty member, because Tom had been impressed by what he heard about him and because I wanted to avoid working for one of our friends. Of course, because of my status as faculty wife, Gerry was soon also a good friend.

Working with Gerry shaped my development in crucial ways. He had been well-trained as a sociologist at New York University and was instrumental in exposing me to a thoroughly sociological perspective. His expertise and enthusiasms also contributed importantly to my subsequent research. He had been working in the areas of medical sociology and the sociology of science. He had also recently published a general theory of organizations in *ASQ*. I was interested in studying universities as organizations, and his expertise fit well with that interest.

Gerry was very much the academic entrepreneur. He always had big grants while at Cornell, and he was able to get me smaller sums of money for my own research. I started out assisting him on one of his projects, but he soon encouraged me to write a proposal for research on universities. What was especially remarkable was that he explicitly encouraged me to write a proposal for the research I thought should be done on universities without worrying about what it might cost. By expressing such confidence in me, he gave me confidence in myself. I began the project with some doubts but soon knew what I wanted to propose. Within a semester, I had developed the proposal that informed both my Masters' thesis and my Ph.D. dissertation. Gerry then went out and raised the money and other assistance I needed to do the research, which consisted of a large-scale questionnaire study of all of the faculty in 80 university departments in the physical and social sciences. I spent the early spring preparing and pretesting the questionnaire. I sent out about 3000 individually addressed questionnaires in the spring of 1968, while students were rioting at places like Columbia and Berkeley. Tom and the girls helped me stuff the envelopes.

By this time, Tom had earned a sabbatical leave, and we decided we would like to go to England for the 1968-69 academic year. He was also interested in studying universities and decided to study the new so-called plate-glass universities there. His idea, very innovative for that time, was to study the founding of new organizations. We decided to live in Brighton, near the University of Sussex, which was the oldest and best-established of the plate-glass universities. We knew one faculty member there who had visited Cornell. Also, Sussex had an internationally known Science Policy Unit, with which I hoped to affiliate while we were there. I would take my data with me and would do my data analyses there.

Things turned out pretty much as planned. I was given some kind of courtesy visiting slot with the Science Policy Unit and Tom's research proceeded very well. We lived in a huge Regency period house near the sea that we rented from a member of the British foreign service. The girls attended a little private school run by two elderly sisters that we had selected largely because it had a horseback riding program as an integral part of the curriculum. Both girls became better horsewomen that year and enjoyed riding across the downs, as the hills around Brighton were called, with their classmates. They were less happy about how cold it tended to be in the classrooms, for the school lacked central heating. When they came home from school and joined me for tea in our large kitchen, their cheeks would turn bright red from the warmth.

What didn't work out was my plan to analyze the data I had collected. Somehow, I don't remember how, I had managed to get much of the data punched onto IBM cards before I left. We hired the key punch operators at the computer center to both punch and verify the data. Because of the short time period between the data collection and our leaving for England, Gerry and I had decided not to try to get all of the data in the questionnaires on cards before I left. So I brought only a partial data set with me, on boxes and boxes of cards. (We had managed to bring lots of books and household possession with us because we traveled to England by boat, not plane.) Nevertheless, the data set was huge for those times—probably around 200 variables for over 1800 respondents—and the University of Sussex apparently had a relatively small computer and lacked any canned computer programs with which to analyze it. The Science Policy Unit put me in contact with computer programmers, but they made little progress on the analysis while I was there. I finally gave up and helped Tom with his research by helping him prepare a questionnaire to follow-up on ideas he had gained through interviews and observations. His planned work done, we left Brighton earlier than planned and traveled for two months around Europe with the girls, now 8 and 13, in our new Citroen, which we picked up at the factory outside Paris. During that time, armed student militants took over the Cornell Student Union. A photograph of the students and their guns made the cover of *Time*. As we traveled through Europe and met with various academics there, the main topic of conversation was the student uprisings that were happening throughout the world. I was genuinely distressed, for I had always believed in the importance of the university as a social institution, and I could not help but feel that what was happening would be damaging to that institution in the long run.

We returned to Brighton after our Continental travels to pick up the last of the responses to Tom's survey, packed up the house, and returned to the United States on the *U.S.S. France*. We were all very happy to be back in Ithaca, but Tom and I soon realized that we had come back to a deeply divided campus. The divisions separated not only departments and individuals we knew from each other; some of the married couples we knew suddenly found themselves at odds with each other over the political and social issues embedded in the student protests and what should be done about them. The climate was so charged with ideology and it could not be ignored. Because we had not been part of the proceeding year's turmoil and not expressed our own views, Tom and I were chosen by the Board of Trustees to do a survey of all of the stakeholder groups on the campus.

This role of the trustworthy citizen was not new to either Tom or I. In particular, he had been Chairman of the Board of Cornell United Religious

Work. I had done my part in the community as Chair of the Religious Education Committee at the Unitarian church, in the League of Women Voters, and in the Parent-Teachers Association. I don't remember the results of the survey, but we put off our own work to develop an instrument, collect data, and analyze it. Somehow we carried off the whole exercise to everyone's apparent satisfaction.

Since I had not been able to analyze my data in England, I still had that to do before I could write my Master's thesis. Even at Cornell, the canned data programs did not meet my needs. So I learned some FORTRAN programming, hired a programmer to do the more difficult programming involved in creating some of my measures of consensus and prestige, and took a part-time job during the academic year of 1969-1970 for a committee whose cause I believed in. Its mission was to integrate graduate work in psychology. Cornell had recently formed a campus-wide program in biology to integrate the work being done in the Colleges of Arts and Sciences and Agriculture, and the administration thought that the social sciences might profit from similar integration. The committee was chaired by George Suci, a psychologist from the child development department and another good friend. He explained his strategy to me and I thought it was very smart. He would put off the procedural issues of how students would be supported and how votes would be counted and thus avoid giving the faculty excuses to reject what he saw as the main issue, which was whether psychologists from departments other than psychology could belong to the graduate field of psychology. I helped by bringing people with similar interests in the field together for lunches and meetings and by compiling and issuing a list of all graduate psychology courses offered across the campus. George's strategy was successful. For the first time all of the faculty who had earned Ph.D. degrees in psychology from various departments across the Cornell campus were admitted to the graduate field of their discipline. This was no mean feat, because it meant that any of them could chair a committee for a Ph.D. degree in psychology.

Meanwhile, my work on the thesis proceeded along with little difficulty. I had decided to test the hypotheses on paradigm development for the thesis and this proved to be a good decision. The results supported the hypotheses sufficiently for me not only to complete and defend my thesis successfully by the spring of 1970, but also to subsequently gain acceptance for an article based on them by the *American Sociological Review* (Lodahl and Gordon, 1972). Gerry, who was my thesis chairman, wanted me to submit the article elsewhere, but perhaps because of my status as a faculty wife, I didn't feel I had to follow his advice. When *ASR* asked for revisions rather than rejecting it, I took it in stride. I had submitted the article there because I believed it showed something important. I had to sacrifice some references to Perrow's work on technology as knowledge in the revision process and generally position the article more from a sociology of science than an organizational perspective in order to satisfy the editor and reviewers. At the time, I was not about to argue such points. But I think they were unfortunate, for the published article did not reveal all of the reasoning behind the study. It took two revisions for the article to be accepted.

By that time, the significance of what was happening sunk in. I was elated. Especially satisfying was that it was my decision, my writing, and largely my ideas that had been validated by the acceptance. Gerry had, of course, contributed his expertise in many ways, especially by acquainting me with Thomas Kuhn's seminal book, *The Structure of Scientific Revolutions*, that informed my work, and also by telling me about the Cartter report on departmental reputation, which I used as a sampling frame for my study. Tom had also contributed his advice and support on many matters, large and small. But I had done the thinking and writing work largely by myself. An additional fillip to my feelings of validation was that I had, at that point, never taken a sociology course as such, yet I had won publication in the most prestigious journal in sociology.

This whole experience reinforced views absorbed from Gerry, my husband Tom, and from reading Kuhn that scientific break-throughs were more likely to occur among those who were somewhat marginal to established fields. I probably would not have had tendencies to accept scientific orthodoxies anyway—having so painfully lost my faith in the doctrines of the Catholic religion in which I was raised only a few years before—but my early success as a neophyte to sociological thinking made me convinced that heavy indoctrination in one scientific perspective does not necessarily produce the best science.

TRIUMPH AND TRAGEDY

While I was quietly savoring my triumph, for it would have been very unseemly for me to let these feelings out to anyone at that time, Tom was going through a period of personal crisis. He didn't talk about it, but I knew something was wrong. I remember being concerned and reassuring myself that everyone goes through crises, and that in the end, he would be fine. I had such confidence in him as a person. I could not imagine that he would not be able to surmount any difficulties.

After receiving my M.S. degree, I was easily persuaded by faculty friends in the ILR School that I should continue my graduate work and pursue a Ph.D. At this point Tom acted totally supportive of my plans. So I took another two years of coursework, including my first courses in the sociology department, and continued to analyze the data on university departments. By this time, the SPSS statistical package had come out, and the computer programmers at the computation center used by the social sciences debugged the programs using my data set. That was great for me, because it meant that my data were on-line in disk storage, rather than on a tape that had to be mounted. There was lots of data left to analyze.

One of those years Gerry went on a sabbatical leave and I spent the time he was gone occupying his office, analyzing piles of data, and writing up the results. He had raised enough money so I could hire a part-time assistant. I trained her in some SPSS procedures and every day we submitted and produced additional analyses. Before the year was out most of the data were thoroughly analyzed. I began writing up some of the results for publication, but saved the data on power issues for my dissertation.

But before I could write the dissertation, another big hurdle remained. I had to pass the comprehensive examination. This was a big bugaboo for me, perhaps because Tom had been flunked on his first comprehensive exam. I hated the idea of an oral exam. I had only taken one before in my life—for honors in music at Wisconsin—and felt I had done miserably in it. But I knew that I somehow would get over this hurdle, too, and I went ahead and scheduled the exam during the spring of 1972. To add to my tensions, Tom had been behaving more and more coldly toward me since sometime that winter, when we had tried revising a joint paper we had written earlier to meet the comments of an editor. The editor had unfortunately said he liked my data but not Tom's. I had insisted on keeping his data in the paper, but the matter had never been satisfactorily resolved.

When the day of my comprehensive exam came, however, Tom seemed to be his usual concerned, sympathetic, and supportive self. He took me out to lunch, and urged me to relax. He often thought I was too "up-tight." So I had a glass of wine with lunch, took a tranquilizer, and went off to take the exam, which was scheduled for 1 p.m. It's a wonder I didn't pass out. Instead I was passed. Gerry Gordon remarked later, "You were great. You just answered the questions and didn't go on and on." I guess he thought I usually did.

My big hurdle was over, and I could now proceed to work on the dissertation, I thought. But matters at home grew more and more strained. I had passed the exam in April or early May of 1972. In early July Tom told me he was leaving the marriage and moved out of the house. He didn't tell me why. I was devastated and lost 20 pounds by the end of the summer. I was also scared—I had built my whole future around Tom and his career. I had not prepared myself mentally or practically for a professorial career, but for a research career. I had envisioned myself getting grants to do research full-time. Clearly, I would now have to take on the full professorial role to support myself and the children.

Subsequent events suggest that, somewhere along the line, perhaps while we in England, perhaps afterwards, Tom had begun to grow alienated from the professorial career and even from our way of life. After about a dozen years on the faculty and soon after our marriage ended, he left Cornell to become a full-time consultant—a role he continues to this day.

In the summer the marriage ended, Claire, after some hesitation, decided to go ahead and attend Cornell as a freshman in the fall. As I had intended all along, she would move into a dorm so she would be as free to pursue her own inclinations as I had been as a student in Madison. But meanwhile, over the summer, she joined a stable in another town so she could show her new thoroughbred horse in class-A horse shows. Andrea, who was 12, was in a cast with a broken leg. I learned to load her horse in our horse trailer and went with her to as many horse shows as I could that summer. She rode with the foot in a cast in a large man's stirrup. She was scheduled to begin junior high in the fall.

By the fall of 1972, I had sold the family home, moved to an apartment, and recovered sufficient equanimity to resume my own studies. But I couldn't quite face the dissertation. At the time I felt getting a Ph.D. degree had cost me my marriage. Fortunately, other opportunities to use my research skills came my way to keep me busy. I was approached by a faculty member in the nutrition school to help design a study. Just before Christmas, Harrison Trice, a faculty member at the ILR School, phoned and asked me if I was interested in advising him on a new research project. He knew that I had been working with Gerry on issues of organizational structure and he needed someone with that expertise, he said. I jumped at the chance. Not only would it give me something to do, but it sounded like exactly the kind of group project that I had always wanted to work on.

While I had never taken a course from Harry, we had come to know one another slightly from various encounters over the years. Then sometime in November of 1972, we had met in the waiting room of our respective doctor's offices one afternoon. In a brief conversation we had there, we learned that we had both been seeking treatment for stress brought on by the breakups of our marriages. We agreed that dealing with lawyers was very stressful and agreed we would have lunch some time and talk more about it. We ended up having lunch together every Monday for months after that. They were hurried lunches because of Harry's teaching schedule, and the conversations were indeed often about difficulties we were having with the lawyers handling our divorces. When Harry called the next month to ask me to join his project, I did not suspect that he might have had personal as well as professional motives in asking. If he did, he never told me, and I never asked. In any case, one very cold January afternoon, after attending one of the meetings of his research group, I found my car would not start. He got the car jump-started, gave me a ride home from the garage where I was having the battery charged, and then asked me to dinner. I accepted, and not long thereafter, our professional relationship had developed into a romance.

Harry's caring and admiration were like a tonic. Tom's leaving had left my already bruised ego severely wounded. Harry made me feel like the "wonderful woman" he often called me. We had many interests in common—he had attended UW at Madison as a graduate student at the same time I was an undergraduate there. His two children were not much older than Claire. Our political and personal values were very similar. But above all, we very much enjoyed talking to each other. We never were at a loss for something interesting to talk and laugh about. My children remarked to me, "It's wonderful how you laugh together." With Harry's emotional support, I tackled my dissertation, writing one chapter a week. By April, I had successfully defended it.

Meanwhile I was applying for professorial positions of several types. I thought I could either teach in an organizational behavior group in a business or management school or else take a position in higher education in a school of education. Both Gerry and Harry would occasionally ask if I really wanted to leave Ithaca, and I would always emphatically reply that I did. But my search wasn't turning up much. It wasn't until late in the winter that I learned that Gerry was writing letters of recommendation that either prominently mentioned my divorce from Tom or said so little about me as to appear unfavorable. I confronted him about it and he acted innocent and promised to do better. I had been unwise enough to tell him that I would stay at Cornell and work on a book with him if I didn't find a job I wanted. He may have believed, as he told me at one point, that as a middle-aged woman I couldn't expect to get a really good job. I was 38 at the time. Finally, after the confrontation and probably through the good offices of Larry Williams, who was on my Ph.D. committee, I got one job interview. It was at the Management School of the State University of New York at Buffalo. I went on the interview and was offered the job. I stalled initially because I still hoped to have other possibilities. But when nothing else surfaced, I finally accepted in March. The department there was clearly research-oriented and had a Ph.D. program, so my major criteria were satisfied. It wasn't until April that another possibility turned up—the Education School at the University of California in Berkeley contacted me about a position in higher education. The person who phoned me was upset to learn I had already accepted a position elsewhere and would not consider changing my mind. Much later I learned that junior faculty in at least one other school had tried to recruit me, but were told by their senior faculty that they would not consider hiring a divorced woman because she would break up marriages. I obviously had several counts against me in the job market of the time.

After I had accepted the position in Buffalo, Harry began to inquire plaintively whether I wouldn't consider coming back. I told him that I would not. After what I had been through I felt I would never again do anything that would put my career second. I knew that, as a Cornell graduate, my chances of being hired to a permanent position at Cornell were poor, and I would not compromise my research interests by joining the faculty at any of the colleges in the Ithaca area.

So I went off to Buffalo with determination to begin a new life and a serious career. I would earn my own way and never have to depend on anyone else again for financial support. More important, I would finally find out what I could do on my own and perhaps find some measure of personal fulfillment. That summer, while we were in New York City attending the American Sociological Association (ASA) annual meeting together, Harry proposed. I told him it was too early for me to consider making such a commitment. I needed to see what my new life would be like. As it turned out, he made sure he was an important part of the new life, coming to Buffalo weekly and sometimes oftener to see me, and phoning at least daily. By Christmas time I gave in to what then seemed inevitable and accepted his proposal. We were married the following October.

MY COLLABORATION WITH HARRY

From the beginning, Harry and I worked very well together. We each respected each others' ideas and our areas of expertise and our talents were complementary. Harry did most of the literature reviews. I did most of the data analysis. We both wrote, but I often did the polishing. The same held true for instrument construction. Harry and the people he hired to assist him managed the administrative details and the logistics of travel. We both interviewed a lot, but Harry's teaching schedule was usually less demanding than mine, so he did the bulk of the interviewing during the school year. We both had graduate students who participated in our efforts, but Harry usually had the bulk of the funding, and therefore the larger support staff.

We did two large-scale empirical studies together. The first, which began in early 1973, was a study of the implementation of alcoholism policies in the agencies of the federal government. We wrote many articles and our first book, *Implementing Change*, from the first study. Harry was a shrewd but principled academic entrepreneur. He studied alcoholism because he found it interesting as a social problem, but also because he genuinely wanted to help people with drinking problems. I was pleased to be joining him in working in areas that had social and not just academic significance. After he joined the ILR School, Harry focused on studying alcoholism programs in work organizations because he came to believe they could make an important contribution to dealing with alcoholism, and also because he was interested in workplaces and was thus able to collect data on other important work-related issues. Some of our articles had nothing to do with alcoholism and were targeted to academic audiences; for example, we co-authored a paper on organizational commitment with one of my Ph.D. students, Jack Stevens. Other articles were written for journals or books read by practitioners, including medical doctors. We wanted our research findings to be applied. We also presented the results to a wide variety of audiences and testified in Congress about them.

While working on the book from the first study, we wrote a proposal for funding a second. We had obtained entry to study the institutionalization of the alcoholism policy in the Western Electric Company. As we had done in the first study, we spent a year getting to know the company and conducting preliminary interviews. The data collection took most of the second year. Again we collected the data at multiple locations with structured face-to-face interviews with a large sample of managers at all levels in the company. While the data collection phase was very arduous, Harry and I both loved it. We were inherently curious and always found something unexpected—sometimes something amusing or merely interesting, a few times something very tragic. Two times during this study we entered a location soon after a murder had been committed by one of the employees. Both times management was visibly shaken and talked to us in almost a cathartic mode. Two other times we encountered wildcat strikes. Any illusions I might have had about the rationality of workplaces was dispelled during my experiences collecting data in these two studies. I saw at first hand some of the realities of organizational life that are glossed over or ignored in most of our theories and texts. Working with Harry greatly reinforced my sociological leanings and perspective. He had been trained at the University of Wisconsin by a Weberian scholar named Howard Becker and the well-known expert in deviant behavior Marshall Clinard. He had always been an active contributor to sociological meetings and journals. After we began working together we invariably attended and gave papers at both the Eastern and American Sociological meetings together. We also presented papers at the Society for the Study of Social Problems meetings, which precede those of ASA. I met some of my best friends at ASA meetings—notably, Carolyn Dexter and Dick Osborn. I had attended my first ASA meeting before starting to work with Harry, but I probably would not have continued the connection so steadily after going to work in a Management School if I had not been working with him.

Harry and I had very different work styles. I have a strong completion drive. Harry was easily bored and could work on many projects at the same time. At various times I would stop and write down all of the projects we were working on so I could keep track of them. There were usually around ten papers or projects in various stages of completion, including those already submitted that would no doubt need to be revised. Each of us also did some projects and papers without the other. The early years of our marriage were a busy and productive time.

When the concept of culture was rediscovered by organizational scholars in the early 1980s, Harry was more than ready to contribute. He had also studied anthropology at Wisconsin with C.W.M. Hart and had already written an article published in the *Industrial and Labor Relations Review* in 1969 comparing personnel practices to cultural rites. He thus immediately became enthusiastic about doing something more in the area. He also had done considerable qualitative research as a graduate student. As usual, although I was still struggling to analyze data on organizational structure and other issues in which I was interested, his enthusiasms won the day. We were not able to collect data on culture right away, however, because we were still in the middle of various papers from the Western Electric study. So we wrote a number of conceptual pieces on culture, some for journals and some for conferences and book collections.

At about the same time, I was becoming more and more active in professional affairs—serving on a grant review panel for the National Institute for Education, editing a special issue of *ASQ*, and becoming program chair elect for the Organization Theory Division of the Academy of Management. All of these activities brought me into contact with many of the leading academics in the field in situations from which I could learn a lot about the state of the field and about academic life and values. In particular, I became reacquainted with Bill Starbuck and Karl Weick, not as a faculty wife, but as a professional in my own right. I also got to know Karlene Roberts and Gerry Salancik. I enjoyed these activities because they enabled me to get to know the well-known scholars in the field personally and made me feel part of a close community of scholars. As these professional obligations increased, it became harder and harder for me to finish all of the planned data analyses on the Western Electric data. While we published many articles from that study, one of which won a prize from the *Journal of Studies on Alcohol*, we never really finished milking the data and we never wrote the book we had intended to write from the results.

Instead we moved into writing and thinking more about cultures in organizations. We had written a single chapter for a book on culture and signed a contract for it with a publisher when, in 1983, I was asked to be Editor of the *Academy of Management Journal.* I hesitated because I knew it would greatly interrupt my own research. At the same time I was greatly honored. Harry solidified my leanings to accept when he said to me, "You can't refuse a thing like that." So I accepted. I don't think he realized how big an interruption the editorship would be to our work together. If he had taken an editorship, he would probably have been able to do it and continue many of his other commitments at the same time. Not me. I gave it my all because doing less would not satisfy my standards. The story of how I planned and executed my editorship has been told elsewhere, in a chapter entitled "Becoming Editor" (Beyer, forthcoming). During the editorship I found myself too busy and intellectually exhausted to do much writing on my own or with Harry. I didn't mind the interruption too much, but Harry did.

Editing the *Journal* turned out to be the most satisfying thing I have done in my academic career. Although it already had a very good reputation, I felt

I could find ways to improve the *Journal*. Perhaps because of my work in the sociology of science, I also felt that being an editor of a scientific journal was a very serious scientific obligation. For that reason, I thought hard about what I could do as editor that would make the *Journal* serve the field better than before. My predecessor Tom Mahoney suggested I use some kind of assistant editors. I ended up naming three scholars—Rick Mowday, Manny London, and Carl Zeithaml—as consulting editors. They had expertise that was complementary to mine and would help me to make good decisions about a wide range of manuscripts. They and all of the members of the Board were wonderful colleagues. They were very conscientious, thorough, prompt, and generally considerate of our authors. In the end, I felt we had made a difference and that *AMJ*, which was a high quality journal when I assumed the editorship, became even better. I probably also enjoyed the job because it put me into contact with so many other people in the field. I felt an extra jolt of satisfaction when I was able to publish the early work of outstanding scholars like Bob Sutton and Connie Gersick.

MY MARRIAGE TO HARRY

Although, as mentioned earlier, Harry proposed before I joined the faculty at SUNY-AB, I did not immediately accept his proposal. One reason was that I could not see how the marriage would work. I was concerned about the fact that we were not living in the same community and wondered if we would ever be able to get a job together. I did not relish the idea of a commuting marriage. I wanted to be part of a community again—something I felt would not be possible in Ithaca anymore. I wanted to settle down and be part of a new circle of friends. I did not see how the life I wanted would be possible if Harry and I were not living together but traveling back and forth every weekend. But Harry was very persuasive and I loved him deeply. He also made sure he was around a lot. Andrea was still living with me and he argued that I should not move again for her sake until she was through with high school, which was four years away. He also indicated he would then be willing to look for a job together somewhere. So, after a few months of thinking and discussion, I said yes. Harry wanted me to have an engagement ring and together we chose an emerald—my birthstone.

We could not marry until our respective divorces were final. Mine was the more difficult divorce and therefore the longest in coming—it took two years and three attorneys to get a divorce decree two years after Tom had left the marriage. Finally, in the summer of 1974, we were both free. We married the following October on Columbus Day weekend.

The first blow to our marriage came just a month later, when Andrea decided to go live with her father in Ithaca. She missed her life there and her sister, who was attending Cornell. She had never really adjusted to the schools and culture in the Buffalo area. Tom had formed a household in Ithaca with the woman who would be his second wife and her children. Despite my best efforts, I could not compete with the farm, the horses, and the family setting he could offer her. Again I was devastated. I felt rejected, and I was genuinely worried about what Andrea would be exposed to in her father's household. His way of life had become quite bohemian.

Not only was I grief-stricken and Harry frustrated about his inability to completely console me, but the ostensible reasons behind the commuting marriage had been removed. In the following months and years it became increasingly evident that Harry was more wed to Cornell than he was to me. He liked the commuting arrangement, or at least tolerated it very well. I did not. But we managed to be together most weekends, and often spent additional time together at professional meetings or while collecting data. We also spent wonderful vacations together between semesters, usually on one of the islands in the Caribbean. We also talked on the phone every night that we weren't together. Despite not living together on a daily basis, ours was a very close and intense relationship made possible by our immense respect for each other, our intellectual sharing, and our enjoyment of one another's company.

By the mid-1980s, I was growing increasingly restive in Buffalo. I felt out of touch with the main currents in the field, especially the macro side of the field. My department was increasingly micro in focus and the Dean told me flatly that he saw no possibility of hiring anyone else in the macro side of the field. I began to look around at other positions, but nothing especially appealing came up until Bill Starbuck went to New York University and decided he wanted to hire me there. While I was not attracted to the idea of working in New York City, I was attracted to the idea of working with Bill and the other people there. When, in 1986, I was formally given an offer and decided to make the move Harry became very angry. He knew that I was considering the position but didn't really protest until I had accepted the job. He argued it would be a much longer commute. He had always done more of the commuting during the school year than I did and did not relish the prospect of a longer car trip. I did not feel I could promise to do more of the driving and argued instead that he should arrange to spend a semester or more at the New York City office of the Cornell ILR School. He did not even consider the idea. As it turned out, the drive to the house I had purchased in New Jersey only took about a half hour longer than the three hours it had taken to drive from Ithaca to Buffalo, but the question of who would do the driving now became an issue as it never had before. I become more resistant to doing much driving because the trip out of the metropolitan area on a Friday and the return home on Sunday was through very heavy traffic for me, while Harry could drive against the traffic. I also felt his schedule was easier than mine. After commuting by train during the week into New York City, and struggling to keep up with the workload I had as the *AMJ* Editor, I was simply too tired by the weekend to do more commuting. So Harry did most of the driving but with less good will then before. We occasionally compromised by meeting in the Poconos and spending the weekend there.

When I accepted the position at NYU I expected that Harry would retire shortly and join me. Harry must have felt that I was not valuing him and his preferences sufficiently. We began to feel estranged from one another. When we tried working on the book on culture together, as Harry very much wanted me to do, I found myself too spent to be able to contribute as I used to. I had been dealing with so many different sets of ideas in the papers submitted to *AMJ* during the week that I just could not focus on the weekend on one more set of ideas with the same intensity as I had in the past.

The result was that Harry became more and more upset about my editorial and other professional obligations. He argued that I was wasting my abilities by devoting myself so assiduously to trying to improve others' research reports for publication in *AMJ*. He became especially upset when I was elected Program Chair of the Academy. For he knew that it meant that I would not have enough time or energy for several years to work with him as intensely as I used to do. It was hard for me to sympathize with his anger, for he had been able to take the opportunities that came his way without my objection. I felt that some time it should be my turn to have the top priority.

The whole matter came to a head when I was recruited in 1987 for a position at the University of Texas in Austin. Harry did not object to my considering the position, perhaps because I was initially very dubious that I would be interested in moving there. But after I saw Austin and met the people on the faculty, I was sold. Although he was invited along, Harry would not accompany me on my recruiting trips to Austin. After I had formally been offered and accepted the job, he filed for divorce. He did not really want the divorce—I think he wanted me to know how much he was hurt. I knew that he would be required to retire in two years, and didn't understand why we could not continue commuting until then. Alternatively, the people in Austin were willing to try to find a position at UT for Harry. But he was too proud and too comfortable and entrenched in his position at Cornell to consider that option.

When I moved to Austin in 1988, we separated and did not see one another for about a year. Sometime in 1990 we resumed our research and writing collaboration and spent the next year and a half finishing the book on culture. We also recovered some of our personal closeness and love. But we never lived together again. When he came to live in Austin in 1991, after his retirement from Cornell, he bought a separate house. He subsequently resided some of the time in Austin and some of it in Ithaca. When he was in Austin, he was usually impatient that I did not have more time to travel around and do things with him. But we did manage some vacations and short trips, and continued our writing together.

The last time I saw him was when he came to visit me in Hanover while I was visiting at the Amos Tuck School at Dartmouth during the fall term of 1994. We drove around enjoying and taking photos of the fall foliage and visiting the picturesque little towns in the vicinity and nearby in Vermont. We had planned to do a little work together, but he didn't really seem interested. He stayed only a few days. We subsequently talked a lot on the phone and planned to vacation after Christmas in the Cayman Islands.

In mid-December, when I returned to the apartment I was renting, from a trip to the Boston area to collect data, I found a message from Harry's daughter Catherine on the answering machine. With my chest and head pounding, I returned the call. I knew something bad had happened to Harry and imagined him in a hospital. But it was worse than that. It was the worst news imaginable. Harry had been hit by a car and killed while crossing the street in a little town in North Carolina. When I realized what had happened a terrible cry emanated from somewhere deep inside me. I knew it was over. I had spoken to him by phone before he left on that trip. I would never have the chance to speak to him again. The pain and sense of loss was greater than I could ever have imagined. An additional painful fact was that his body had already been cremated when I learned of the death and so I would never have the chance to see him again to say good-bye.

I was touched by the number of letters and other messages of condolence I received over the next few months from people Harry and I have known. I held a memorial service in Austin with my close friends to help me to say the good-bye I felt I needed to say. Another important part of my life was ended.

ON MY OWN

I brought many high hopes and heavy obligations with me to Austin. I had just been elected Program Chair of the Academy and had to somehow organize my office and some supporting staff to carry on those obligations. In addition, I wanted to help to build a strong department into an even better one. My colleagues and the Dean were very supportive to both sets of efforts. I was able to get some time off from teaching and enough support staff to help me with the many tasks involved in the succession of offices from Program Chair to President. Lisa Berlinger, a Ph.D. student at the time, was especially helpful in putting the program together for the 1989 Academy meetings in Washington. She and I became fast friends from that time on, and I later served on her dissertation committee. During the first two years I was in Austin I also headed up the organizations group's recruiting effort, by which we managed to hire two outstanding additions to the faculty—Janet Dukerich and Allison Davis Blake.

Of the various offices I held in the Academy, that of Program Chair was most satisfying. I was proud of the program I put together on the Social Consequences of Management, and really enjoyed working with various individuals who contributed their efforts to one or another part of the program. Jeff Sonnenfeld, Carolyn Dexter, and Peter Frost were especially helpful in obtaining a stellar slate of speakers for All-Academy events. The year as President was less satisfying. Although I did manage to realize some of the aims I had—particularly establishing a strategic planning committee—I was frustrated in not being able to obtain support from the Board of Governors for others—founding a new journal on the discipline itself and reining in the increasing differentiation of the Academy into new divisions. When it became obvious there was little support for a new journal, I pushed for and won support for additional issues of *AMJ*. In regard to the differentiation issue, most of the Board members felt that the founding of new divisions was a necessary and natural response to the demands of our environment and our members. While I agreed, I was concerned that members of the Academy would begin to feel their main allegiance to their divisions and that the organization as a whole would fragment into groups that were relatively isolated from each other. My efforts did lead to an increased awareness that we should build some mechanisms for integration, and that is now being done at the same time as divisions are being encouraged to have more activities on their own. Whether the integrative devices being used are enough to counter the centrifugal forces of specialized identities remains to be seen.

Meanwhile, getting my own research efforts underway again was proving to be very difficult. I had brought some data with me from NYU on what had happened during the review process during my term as Editor of *AMJ*. Most of the data had already been already coded and put in an SPSS file by Bill Fox when I was at NYU, but there was some additional interpretive coding to do. Lisa Berlinger helped with that part of the project, but moved on to do her dissertation. It wasn't until a student from sociology, Roland Chanove, decided to transfer into our Ph.D. program that I was able to find a student with relevant skills and interests to work with on this project. He and I worked with the data for two years, but the analysis involved a lot of complicated procedures, and it proceeded slowly. Unfortunately for both of us, it turned out that Roland didn't really fit very well with expectations in the management department, and he returned to the sociology department. I was left to finish the paper and subsequent analyses more or less on my own, which I managed to find time to do eventually (Beyer, Chanove, and Fox, 1995). Other projects were begun with other students, but because of the department's policy to reassign students to work with different faculty every year, I found it difficult to finish them.

Because my research interests had turned to culture, I embarked on doing qualitative research after I came to UT. It has turned out to be much more time-consuming and difficult than the quantitative research I have done in the past. While I have always chosen to focus most of my attention on issues that were new or neglected in past research, I have now added an additional criterion: that what I study have general social significance. With this new criterion in mind, I have started projects on childcare for working parents, which I see as vital to the welfare of families; on the SEMATECH consortium, which I considered an important social experiment in interfirm cooperation; and on total quality management, which I see as a potentially revolutionary change in management practice. I have collected lots of qualitative data on all of them but, on my own, have had great difficulty completing the data analyses. As in many other schools, management students who might want to work with me find it hard to acquire the skills and logic of qualitative methods at the same time as they are being asked to master quantitative ones. But I continue to plug away on with these projects and am finally coming close to fruition on two of them, thanks to the cumulative assistance over the years of Audrey Chia, Mason Carpenter, Chris Edwards, David Hannah, and Frances Hauge. At the same time, I have been kept busy advising and learning with many different Ph.D. students working on their dissertations—among them Ana Adinaya, Lisa Berlinger, Dulce Pugliese, Steve Lutze, Wanda Mohr, dt ogilvie, Amy Pablo, Pam Ryan, Lynn Scott, and Kathie Sutcliffe. In several cases, close continuing personal and intellectual relationships have grown out of these contacts.

Another factor that has delayed completion of my research projects is that I am asked to do a wide variety of theoretical papers and presentations that are hard to turn down because they follow up on work I have done in the past. The debate at the 1995 Academy meetings on paradigms in management is an example of such a tempting opportunity. Another is working with a committee for the National Research Council to prepare a chapter on how culture is related to human performance. While I don't accept every such opportunity by any means, fulfilling the ones I do accept takes time and attention away from my ongoing research.

My continuing to hold various professional offices has undoubtedly also interfered with progress on my research. Most recently I served as President of the International Federation of Scholarly Associations in Management—an umbrella organization of professional associations from various countries to which the Academy of Management belongs. As the second president of that organization I spent many hours struggling to get it organized. Differences in academic customs in various countries and other cultural differences made the job extremely difficult at times. In addition I have been fairly active off and on in the Society for the Advancement of Socio-Economics (SASE) and some of the organizations of faculty women on the UT campus. All of these are groups that I want to support because I believe in what they are trying to do, but the obligations they entail further fragment my time.

Because I have a strong completion drive and was accustomed to being able to finish things, I find my recent difficulties in bringing my research to fruition very frustrating. I have to keep reminding myself that I did manage to complete some projects in the last few years. I finished the book with Harry. I finally finished the paper on the *AMJ* review process, which entailed getting through a tough review process that required me to reframe the study and rerun all of the statistical analyses. I've written a number of book chapters, including this one. Happily, I've also found new collaborators.

Shortly after I joined the UT faculty Larry Browning, a faculty member in the Communications Department, sought me out and pointed out the similarities of our interests. Larry was very warm and generous to me; he offered both his friendship and his extensive experience doing qualitative research. He taught some of my students qualitative methods and sent his students to me for my courses on culture. We served on various Ph.D. committees together with both my students and his. Initially we didn't find the time or occasion to do collaborative research, but when it turned out that we were both collecting data on SEMATECH, we finally got together and pooled our data for a collaborative effort (Browning, Beyer, and Shetler, 1995). Working with Larry on that paper was stimulating and fun. Our skills and viewpoints are complementary, and our ideas just sparked off one another. We have another paper in the revise and resubmit stage that we're determined to find the time to finish soon.

When, in connection with my research at SEMATECH, I was given an opportunity to study total quality management in some of the supplier firms in the semiconductor industry, I knew I needed another experienced researcher to help me to collect data. As it happened, I had been in frequent contact with Dick Osborn at the same time because he was the Program Chair for the 1994 IFSAM biennial meeting in Dallas. From conversations we had had, I knew he has also been doing research in high tech firms. I now asked him to work with me on the TQM project, and he agreed. We put together an instrument and spent several pleasant but demanding weeks collecting data at two firms on the East and West coasts. We have recently gone back to collect a second round of data on these firms, but have not yet found the time to work together on analyzing all of the data. It's another project I'll have to find the time for.

Now that my heavy period of involvement with the Academy and IFSAM are over, I feel I can devote more time to renewing my basic disciplinary orientation in sociology. Two factors encouraging that effort have been that the sociology department at UT extended a joint appointment to me several years ago and that Amitai Etzioni and Nancy di Tomaso have encouraged me to become active in SASE. I have managed to attend several SASE meetings and thus to renew ties with like-minded sociologists. I was also asked to help organize the program in organizations for the 1996 ASA meetings. After my many years of intense service in the field of management, being able to find time for sociology meetings seems a luxury.

It strikes me that what I have experienced and am currently experiencing is not unique. As our careers progress we accumulate a wide variety of obligations and interests. If we don't also acquire some substantial amount of good assistance or collaboration, our research is bound to suffer. It was risky for me to begin a whole new line of research using methods that I had to learn at this point in my career. But I have been converted to the cultural perspective and am determined to do research consistent with that perspective. And I've always been attracted to new challenges.

Right now, the achieving is more difficult than it used to be and the groups to which I feel I belong are more geographically scattered. I have many real friends in the field with whom I keep in touch and to whom I know I can turn for practical or moral support. One of the clear benefits of holding the offices I have held in the Academy and other organizations is the people I have been privileged to get to know and the many close friends I have made in the process. My family, however, is scattered and it's hard for me to find time to see them as often as I'd like. Claire, now divorced and a practicing veterinarian, has three daughters and lives in Cortez, Colorado. Andrea, who became a lawyer and married another, works for the Swiss Bank, has a son and daughter, and lives in New Jersey. Both of them have busy and demanding careers. We can rarely all get together as a family. So we try to stay in touch by phone and I stop to visit them when I can between professional trips. Meanwhile my mother is aging and lives alone because my father died more than a decade ago. She's still in Milwaukee, another plane trip away.

There's always more to do and that I want to do than I can accomplish. While I feel freer to decide for myself what I will try to achieve and what I will decide to ignore than I used to, the choices don't get easier—especially since I have learned the hard way that we cannot foretell the trade-offs that may be involved when we make our choices. I am trying to become more selective now about how I spend my energies and time. I will no doubt continue to become involved in various group efforts because I enjoy the feeling of belonging that comes from working with interesting people toward group goals. At the same time I want to continue to do research and to write, which I enjoy now more than ever. This very afternoon I plan to explore some new research ideas with Donde Ashmos after we've finished our tennis game. We've been trying to make time in our respective schedules to do research together for several years now. And that's not all. Somehow I must also find time and energy to be a worthy daughter, a good mother, a beloved grandmother, a good friend to the many people who have been a good friend to me, and a good citizen. There's a lot of balancing left to do.

As I look back, I can see that performing and achieving were initially routes to belonging for me. As my life progressed the demands connected with performing and achieving seemed to eat away at my closest relationships and I was left without the belonging a marriage can give. I didn't want it to happen that way. But it did. It seems, as I look around at colleagues, that the prices for women's achievements were often higher, at least for my generation, than they were for men doing the same things. At the same time I have been privileged in many ways. When I consider where I started, as writing this chapter encouraged me to do, I recognize how far I have come from my origins. The journey has sometimes been easy and sometimes very hard, but it has never been dull. Most satisfying of all, I have often found the general sense of belonging I was seeking as my efforts brought me many opportunities—not only to perform and achieve, but to come to know and work with many wonderful people. I treasure them all.

PUBLICATIONS

1969

Administrative science quarterly index: Vols 1-12. Ithaca, NY: Cornell University.

With P. Anisef, S. Marquis, E. Morse, L.R. Putnam, J.A. Stewart, & T.M. Lodahl. A review article: The handbook of organizations. *Administrative Science Quarterly, 13*, 320-337.

1972

With G. Gordon. The structure of scientific fields and the functioning of university graduate departments. *American Sociological Review, 37*, 57-72.

1973

With G. Gordon. Funding the sciences in university departments. *Educational Record, 54*, 74-82.

With G. Gordon. Differences between physical and social sciences in university graduate departments. *Research in Higher Education, 1*, 191-213.

1974

With R. Snipper. Objective versus subjective indicators of quality in graduate education. *Sociology of Education, 47*, 541-557.

[Special reviews: T. Parsons & G.M. Platt, *The American university*]. *Journal of Higher Education*, pp. 553-557.

1975

With J. M. Stevens. Unterschiede zwischen einzelnen Wissenschafter im Hinblick auf Forschungasktivat und Produktivatat. In N. Stehr & R. Koenig (Eds.), *Wissenschaftssoziologies: Materialien und Studien* (pp. 349-374). Koeln und Opladen: Westdeutscher Verlag. [Special Issue on the Sociology of Science, *Kolner Zeitschrift fur Sociologie and Sozialpsychologie*].

1976

With T.M. Lodahl. Patterns of influence in U.S. and English universities. *Administrative Science Quarterly, 21*, 104-129.

1977

With H.M. Trice. A sociological property of drugs: Differential acceptance of users of alcohol and other drugs among university undergraduates. *Journal of Studies on Alcohol, 38*, 58-74.

With H.M. Trice & R.E. Hunt. Alcoholism programs in unionized work settings: Problems and prospects in union management cooperation. *Journal of Drug Issues, 7*, 103-115.

With J.M. Stevens. Factors associated with changes in prestige of university departments. *Research in Higher Education, 7*, 229-255.

With H.M. Trice. Differential use of an alcoholism policy in federal organizations by skill level of employees. In C.J. Schramm (Ed.), *Alcoholism and its treatment in industry* (pp. 44-68). Baltimore: Johns Hopkins University Press.

1978

With H.M. Trice. *Implementing change: Alcoholism policies in work organizations.* New York: Free Press.

Editorial policies and practices among leading journals in four scientific fields. *Sociological Quarterly, 19*, 68-88.

With H.M. Trice & R.E. Hunt. Evaluating implementation of a job-based alcoholism policy. *Journal of Studies on Alcohol, 39*, 448-465.

With J.M. Stevens & H.M. Trice. Assessing personal, role, and organizational predictors of managerial commitment. *Academy of Management Journal, 21*, 380-396.

[Review of: G. Salancik & B.M. Staw (Eds.), *New directions in organizational behavior.*] *Administrative Science Quarterly, 23*, 488-495.

1979

With H.M. Trice. A reexamination of the relationship between size and various components of organizational complexity. *Administrative Science Quarterly, 24*, 48-64.

With H.M. Trice. Women employees and job-based alcoholism programs. *Journal of Drug Issues, 9*, 371-385. [Reprinted in J. L. Ford (Ed.). *Women and alcohol.* Dubuque, IA: Kendall-Hunt, 1980.]

1980

With R.E. Hunt & H.M. Trice. The impact of federal sector unions on supervisors' use of personnel policies. *Industrial and Labor Relations Review, 33*, 212-231. [Portions reprinted in J. Fossum (Ed.), *Labor Relations: Development, Structure and Process.* Dallas: Business Publications, Inc., 1982.]

With J.M. Stevens & H.M. Trice. Predicting how federal managers perceive criteria used for their promotion. *Public Administration Review, 40*, 55-66.

With K. Provan & C. Kruytbosch. Environmental linkages and power in resource-dependence relations between organizations. *Administrative Science Quarterly, 25*, 200-225.

With J.M. Stevens & H.M. Trice. Managerial receptivity and implementation of policies. *Journal of Management, 6*, 33-54.

1981

With H.M. Trice. A retrospective study of similarities and differences between men and women employees in a job-based alcoholism program from 1965-1977. *Journal of Drug Issues, 11*, 233-262.

With H.M. Trice & C. Coppess. Sowing seeds of change: How work organizations in New York state responded to occupational program consultants. *Journal of Drug Issues, 11*, 311-336.

With H.M. Trice. A data-based examination of selection bias in the evaluation of a job-based alcoholism program. *Alcoholism: Clinical and Experimental Research, 5*, 489-496.

With H.M. Trice. Managerial ideologies and the use of discipline. *Academy of Management Proceedings*, pp. 259-263.

Ideologies, values, and decision making. In P. Nystrom & W. Starbuck (Eds.), *Handbook of organizational design* (Vol. 2, pp. 166-202). New York: Oxford University Press.

With D.P. Simet, A.H. Hageman, & C.V. Fukami. A holistic, experiential course design suitable for large MBA classes. *Proceedings for the Eighth Annual Organizational Behavior Teaching Conference* (pp. 79-81). Harvard Business School.

1982

With H.M. Trice. Social control in work settings: Using the constructive confrontation strategy with problem-drinking employees. *Journal of Drug Issues, 12*, 21-49.

With H.M. Trice. The utilization process: A conceptual framework and synthesis of empirical findings. *Administrative Science Quarterly, 27*, 591-622.

Explaining an unsurprising demonstration: High rejection rates and scarcity of space. *The Behavioral and Brain Sciences, 5*, 202-203. [Reprinted in S. Harnad (Ed.), *Peer commentary on peer review*. Cambridge: Cambridge University Press, 1982.]

Power dependencies and the distribution of influence in universities. In S.B. Bacharach (Ed.), *Research in the sociology of organizations* (Vol. 1, pp. 167-208). Greenwich, CT: JAI Press.

With H.M. Trice. Job-based alcoholism programs: motivating problem drinkers to rehabilitation. In E.M. Patterson & E. Kaufman (Eds.), *Encyclopedic handbook of alcoholism* (pp. 954-980). New York: Gardner Press.

With H.M. Trice. Design and implementation of job-based alcoholism programs: Constructive confrontation strategies and how they work. In Research Monograph-8, *Occupational alcoholism: A review of research issues* (pp. 181-239). Rockville, MD: U.S. Department of Health and Human Services.

Critique on dynamics of the identification and referral process in work organizations. In Research Monograph-8, *Occupational alcoholism: A review of research issues* (pp. 315-322). Rockville, MD: U.S. Department of Health and Human Services.

With M.L. Cummings & N. Kurtz. Recommendations. In Research Monograph-8, *Occupational alcoholism: A review of research issues* (pp. 323-327). Rockville, MD: U.S. Department of Health and Human Services.

1983

With J.M. Stevens & H.M. Trice. Exploring the black box in research on public policy: The implementing organization. In R.H. Hall & R.E. Quinn (Eds.), *Organizational theory and public policy* (pp. 227-243). Beverly Hills, CA: Sage Publications.

With H.M. Trice. Current and prospective roles for linking organizational researchers and users. In R.H. Kilmann et al. (Eds.), *Producing useful knowledge for organizations* (pp. 675-702). New York: Praeger Publications [Reissued San Francisco: Jossey-Bass, 1994].

With G. Jaskolka & H.M. Trice. Measuring and predicting managerial success. *Eastern Academy of Management Proceedings*, pp. 144-147.

1984

With H.M. Trice. A study of union-management cooperation in a long-standing alcoholism program. *Contemporary Drug Problems–A Law Quarterly, 11*, 295-317.

With H.M. Trice. The best-worst technique for measuring work performance in organizational research. *Journal of Organizational Behaviour and Statistics, 1*, 95-115.

With H.M. Trice. Studying organizational culture through rites and ceremonials. *Academy of Management Review, 19*, 653-669. [Reprinted in P. Gagliardi (Ed.), *Le Imprese come culture* (207-233). Torino: Petrini editore, 1986. Reprinted in S.M. Puffer (Ed.), *Management insights from literature*. Boston: PWS-Kent Publishing Co.]

With H.M. Trice. A field study of the use and perceived effects of discipline in controlling work performance. *Academy of Management Journal, 27*, 743-764.

With H.M. Trice. Work-related outcomes of the constructive-confrontation strategy in a job-based alcoholism program. *Journal of Studies on Alcohol, 45*, 393-404.

With H.M. Trice. Employee assistance programs: Blending practical and humanistic ideologies to assist emotionally disturbed employees. In J.R. Greenley (Ed.), *Research in community and mental health* (Vol. 4, pp. 245-297). Greenwich, CT: JAI Press.

[Review of: W.R. Scott, *Organizations: Rational, natural and open systems*]. *Administrative Science Quarterly, 29*, 134-138.

1985

With G. Jaskolka & H.M. Trice. Measuring and predicting managerial success. *Journal of Vocational Behavior, 26*, 189-205.

With H.M. Trice. Using six organizational rites to change culture. In R.H. Kilmann, M.J. Saxton, R. Serpa, & Associates (Eds.), *Gaining control of the corporate culture* (pp. 370-399). San Francisco: Jossey-Bass.

1986

With H.M. Trice. Charisma and its routinization in two social movement organizations. In B.M. Staw & L.L. Cummings (Eds.), *Research in organizational behavior* (Vol. 8, pp. 113-164). Greenwich, CT: JAI Press. [Reprinted in L.L. Cummings & B.M. Staw (Eds.), *Leadership, participation, and group behavior* (pp. 85-136). Greenwich, CT: JAI Press.]

1987

With H.M. Trice. How an organization's rites reveal its culture. *Organizational Dynamics*, Spring, 5-24.

Managing cultural diversity in a world economy. In G. Westacott, C. Dexter, S. Boukis, & J. Yanouzas (Eds.), *Managing in a global economy II* (pp. 4-9). Eastern Academy of Management Proceedings.

Retrospective comment: The essentials of leadership, Mary Parket Follett. In L.E. Boone & D.D. Bowen (Eds.), *The great writings in management and organizational behavior* (pp. 59-61). New York: Random House.

1988

With R. Dunbar & A. Meyer. Comment: The concept of ideology in organizational analysis. *Academy of Management Review, 13*, 483-489.

With H.M. Trice. The communication of power relations in organizations through cultural rites. In M.O. Jones, M.D. Moore, & R.C. Snyder (Eds.), *Inside organizations: Understanding the human dimension* (pp. 141-157). Beverly Hills, CA: Sage Publications.

[Review of: C.A. Enz, *Shared values and power in the corporate culture*]. *Contemporary Sociology, 17*, 323-324.

1989

Ethical dissent when accountability fails: Committing the truth. [Review of: *The Whistleblowers* by M.P. Glazer & P.M. Glazer]. *Science, 244*, 835-836.

1990

With F.J. Milliken & J.E. Dutton. Understanding organizational adaptation to change: The case of work-family issues. *Human Resources Planning, 13*, 91-108.

The twin dilemmas of commitment and coherence posed by high technology. In L.R. Gomez-Mejia & M.W. Lawless (Eds.), *Organizational issues in high technology management* (pp. 19-36). Greenwich, CT: JAI Press.

[Review of: A. Etzioni, *The moral dimension: Toward a new economics*]. *Academy of Management Review, 15*, 329-331.

1991

With H.M. Trice. Cultural leadership in organizations. *Organization Science, 2*, 149-169.

With A. Chia. Prevention is the best cure for ethical dilemmas. *Academy of Management Executive, 6*, 91-94.

1992

Metaphors, misunderstandings and mischief: A commentary. *Organization Science, 3*, 467-474.

Researchers are not cats; they can survive and succeed by being curious. In P.J. Frost & R.E. Stablein (Eds.), *Doing exemplary organizational research* (pp. 65-72). Newbury Park, CA: Sage Publications.

1993

With H.M. Trice. *The cultures of work organizations* (528 pp.). Englewood Cliffs, NJ: Prentice-Hall.

With S. Lutze. Organizations, values, and decision-making. In C.E. Conrad (Ed.), *The ethical nexus* (pp. 23-45), Norwood, NJ: Publishers.

With H.M. Trice. Capturing the essence of organizational culture. *OBTS News and Commentary, 8*(3), 5-6.

1994

With H.M. Trice. [Review of: M. Alvesson & P.O. Berg, *Corporate culture and organizational symbolism*]. *Journal of Marketing, 58*, 125-127.

The big culture shift: Competitors learning to corporate. *Discovery, 14*(1), 6-13.

1995

With L.D. Browning & J.C. Shetler. Building cooperation in a competitive industry: SEMATECH and the semiconductor Industry. *Academy of Management Journal, 38*, 113-151.

With R.M. Chanove & W.J. Fox. The review process and the fate of manuscripts submitted to *AMJ*. *Academy of Management Journal, 38*, 1219-1260.

With H.M. Trice. Writing organizational tales: *The culture of work organizations*. *Organization Science, 6*, 226-228.

FORTHCOMING

Becoming an editor. In P. Frost & S. Taylor (Eds.), *Rhythms of academic lives*. Newbury Park, CA: Sage Publishing.

Going with the flow. In B.H. Reich & D. Cyr (Eds.), *Scaling the ivory tower: Stories of women in business faculties*. Westport, CT: Greenwood Publishing.

Organizational cultures and faculty motivation. In J. Bess (Ed.), *Teaching well and liking it*. Baltimore, MD: The Johns Hopkins University Press.

A Hopscotch Hike

GEERT HOFSTEDE

ROOTS: 1928-1945

In my 1980 book, *Culture's Consequences*, I wrote in an Appendix on "The Author's Values:"

> Research into values cannot be value-free. In fact, few human activities can be value-free. This book reflects ... between the lines the values of its author ... The origins of my value system, like everyone else's, are found in my national background, social class and family roots, education, and life experience. I was born in the Netherlands in 1928 and lived there until 1971. I was the youngest of three children of a high civil servant. Our family relationships were reasonably harmonious; my father had a modest but fixed income, so that we did not suffer from the 1930s economic crisis. There was enough of everything but not luxury; money was unimportant and rarely spoken of. What was important was knowledge and intellectual exercise, at which we were all quite good. I went to regular state schools and liked them. We lived through the German occupation (1940-1945) without physical suffering but detesting the occupants. I was too young at the time to understand the full scope of the ethical issues involved in Nazism, but I had seen my Jewish schoolmates being deported never to return. Only in the years after 1945 did I fully realize that for five years we had lived under a system in which everything I held for white was called black and vice versa; which made me more conscious of what were **my** values, and that it is sometimes necessary to take explicit positions...

For as long as I can remember I was considered a gifted child, and I had soon learned to take advantage of that reputation. If another child did not

Management Laureates, Volume 4, pages 85-122.

ISBN: 1-55938-730-0

know the answer to a question it was blamed on the child, but in my case it was blamed on the question. At primary school they made me jump two classes, so that I entered secondary school (the *Gymnasium*) at the age of ten whereas most others were twelve. Fortunately there were some compensating factors protecting me against becoming too impossible. I spent a lot of my free time with a neighbor family with two boys of my age, where the emphasis was not on intellect but on warmth, fun, and creative games. In 1937, after the World Jamboree came to Holland, I became a Boy Scout, learning lots of practical things. And I got private lessons in handicraft and music, things not taught at school in those days.

The *Gymnasium* was a school for the intellectual elite—not the money elite, for education was free or almost free. The fixed curriculum included six languages. Next to our native Dutch, I got French already in the last two years of the primary school; Latin started in the first *Gymnasium* year, classic Greek and German started in the second, and English, considered the easiest, in the third. The English classes were very modern. We learned pronunciation from linguaphone records played in class, and we were expected to speak only English; there was a fine of one cent for everyone who spoke Dutch, and once a year the proceeds were spent on a party. We also got history, geography, biology, drawing, gymnastics, algebra, geometry, physics and chemistry; all of them compulsory, there were no electives. I don't know how they made it all fit into the program. I loved languages. My favorite classic author was Homer; we read parts of the *Odyssey* in Greek, and I identified with Odysseus as he travelled around the Mediterranean.

The German occupation in the beginning did not interfere much with our lives, but within a year we were strongly affected by it. Food was rationed; borders were closed; the Scouting movement was forbidden; a beloved Jewish teacher was fired and later deported; Jewish children followed. One student suddenly disappeared; there were rumors that he had been in the resistance movement. Proclamations appeared in the newspapers listing the names of university students, some of whom we knew, who had been condemned to death and executed by the Germans for resistance. There were one or two children in my class from Nazi parents, and we stopped talking to them—they became outcasts as long as the war lasted. One day in 1942 a teacher read to us an instruction from the authorities to cut the names of Jewish authors out of our textbooks (we continued using the "authorless" textbooks!). I was thirteen at the time and I did not realize how paranoid a system must be that issues such instructions. A perceptive person could have predicted the Holocaust at that moment.

In my fourth *Gymnasium* year, in 1943, all government ministries were moved from The Hague to various small places up-country; we had to move to the province town of Apeldoorn, so I changed schools. After D-day in June, 1944 the schools closed completely, as the Germans tried to round up the teachers and older students for labor in Germany. I was just too young for that; I spent the fall of 1944 working as a messenger boy and as a cabinet maker's apprentice, and made trips to the countryside on a ramshackle bike trying to buy scarce food from the farmers. I also read a lot: whatever books I could lay my hands on in Dutch, French, English and even in German though the language was not popular.

We were liberated by Canadian troops on April 17, 1945; I will never forget the date. I was back to school in early May, and our government granted all students expecting to graduate in our year the secondary school certificate in June—in spite of the lost year. Neither the usual national examinations nor local tests could be organized at such short notice.

BECOMING SOMEBODY: 1945-1963

The postwar decades were a period of reconstruction and optimism. In hindsight, my generation was blessed to reach adulthood just at such a time. There was a sense of purpose in the air.

The undeserved school certificate gave access to a university study, but my father, who had been a teacher before, strongly advised me to wait a bit and do something else first; the universities were bound to be in a state of chaos for some time, and I was only sixteen. So I joined a technical college of sub-university level (*HTS*) that would give me a certificate in mechanical engineering in three years. Somehow we had always agreed that I should be an engineer. My father had a university degree in engineering; my mother had been a medical analyst. My older sister and brother both had evident literary talents, but no technical interest at all. I, however, had tinkered a lot with a Meccano construction kit, built radios, experimented with chemicals, and was good at using tools.

The *HTS* curriculum provided for one year at school, one year in internships in industry, and a final year in school again. I did the first year, in which I learned lots of practical things like technical drawing and metal working. Then I went into internships. I worked in the fitting shop of a small firm building bakery machines, and in the pattern-making and foundry departments of a large firm building marine diesel engines. I was sent to a marine breaking-up yard to make drawings of an engine salvaged from a war wreck. And during the last months I got a coveted position as an assistant engineer on a steamship making a trip from Amsterdam to Indonesia and back. This was 1947; we took Dutch soldiers to the East and Dutch civilians, released from the Japanese concentration camps, home to Holland. Apart from some family vacations in Belgium and Luxemburg before the war, this was my first international experience, and I loved it.

When I returned in August, 1947 the government had just issued new rules for military conscription. Those having started an academic study before the year of their twentieth birthday would get a deferment; those who were older would first have to serve in Indonesia, for three years. My father advised me to drop my last *HTS* year and to enter university right away. In September, 1947 I was a student at Delft Technical University, where I would graduate in Mechanical Engineering in 1953.

Because of my *HTS* and internships detour I found myself in Delft amidst students of my own age. It was a relief to no longer be the youngest. I became a member of the student *Corps* that maintained strong traditions, including long drinking nights at the *Corps* club and periodic stylish dances, where young ladies wore evening dresses and we young men wore tuxedos. The Corps was the source of life-long friendships. That year we celebrated its 100th anniversary, and we had a week of gorgeous celebrations in July, 1948. Among other things, an open-air theatre had been built on the historical market square of Delft where we performed a musical. I played one of the leading parts in it, which was an unforgettable experience. My partner during the anniversary celebrations was a very lovely English girl, whom I had met the year before when she visited our country with her parents. Later on I stayed with them in London. We kept corresponding, but she was too English and I was too Dutch for complete harmony. In hindsight, this was my first exposure to the pitfalls of cross-cultural communication. Our friendship did, however, improve my English a lot.

Thanks to my *HTS* experience I moved fast through my academic studies at first; only in later years did I go more slowly, mainly because of the non-technical interests I had developed. For several years I served on the editorial team of the Corps journal *De Spiegel* (The Mirror), called an "a-periodical" because its dates of appearance were unpredictable. I wrote poems and fancy stories for it. My poems were ingenious but rarely very poetic, tinkering with words rather than inspired by the Muses. My prose was better.

In 1950 I got a letter from the Royal Dutch Shell Oil Corporation that I was among the (British and Dutch) laureates for a student award (the *Bataafsche Studieprijs*) which implied an annual stipend of two thousand guilders—then a sizable amount—for the rest of my years at university, to be used for extracurricular activities only; I had to write a report to the company each year on how I used it. In the Shell philosophy, extracurricular activities of students were at least as important for future leadership as the university curriculum itself.

The Shell money allowed me, among other things, to serve for a year as the national president of the Liberal Student Christian Association (*VCSB*). While still at the *HTS* I had been confirmed as a member of the Dutch Reformed Church, but my interest in religious matters had really only been developed at university. One nice thing about the *VCSB* job was that it meant visiting other universities; I had discovered at that time that there is more to life than engineering. The other nice thing was that one met female colleagues who were definitely scarce at our technical university. Maaike Van den Hoek was a student of French language and literature at the University of Amsterdam; we fell in love in 1952 and married in 1955.

In my fifth year at Delft I joined an elective workshop on management issues (*Economisch Privatissimum*) led by Professor J.A. Veraart, a nationally famous political economist. In teams of two we had to select a book from a list of recent publications, and to write an essay on it. My colleague and I chose a recent dissertation from a Dutch Jesuit priest, A.M. Kuylaars.[1] Its Dutch title translates into something like "Work and Life of the Industrial Worker as an Object of a Socially Responsible Company Policy." Kuylaars studied the impact of the job content on the life of workers. He was inspired by, among others, the French sociologist Georges Friedmann.[2] Neither Kuylaars nor Friedmann ever became known in the United States, although Kuylaars's research led to similar conclusions as the U.S. study by Frederick Herzberg and his collaborators eight years later.[3] Kuylaars distinguished between the "external" and the "internal" productivity of labor. The external productivity is based on the amount of goods and services produced. The internal productivity is related to human development and self-actualization: Kuylaars used in Dutch the word *zelfverwerkelijking*, the exact equivalent of the term later to be popularized in the United States by Abraham Maslow.[4]

Kuylaars' book was a revelation to me. It allowed me to integrate technology and humanity. Technology was the field I was being trained in; humanity was my deepest concern as a person, and a major theme in the programs of the *VCSB* to which I was strongly committed. The book put me on a track that I never left since.

In the meantime I graduated, in 1953, on a purely technical subject: the design of a high-speed diesel engine. Immediately after graduation I was conscripted into the Dutch army to fulfil my deferred national service. As a graduate engineer I got a fast track training, and spent most of my 20 months as a technical officer in the army purchasing service. I had plenty of time to think about my own life, and one decision I arrived at was that before I would take an engineer's job I wanted to experience for some time the situation of an ordinary worker—not, like during my internships, as a trainee but incognito, on the same terms as other workers.

For four months, from March to July 1955, I worked as a semi-skilled fitter for a printing press manufacturing company in the industrial Northern borough of Amsterdam; then, for another month, I was an unskilled laborer at a cigarette factory. Extracts from my diary were published in Dutch in 1978 and in English in my 1994 book *Uncommon Sense about Organizations*. My brother, who had become a sociologist, had introduced me to his distinguished sociology professor at Leiden university, Fré Van Heek, who encouraged my plans and gave me some important advice and literature about the participant observation method.

These months provided invaluable learning about organizations. For example, it showed me the irrelevance and invisibility of higher management at the worker level. Our immediate boss, a charge hand (crew leader) was our only source of authority; even the next higher man, the fitting shop foreman, rarely interfered. Organizations look very differently from below than they do from above, and those who aspire to higher positions should be aware of this fact—if they ever want to understand how jobs really get done.

Another amazing aspect was the way the incentive payments worked. We were paid a straight piece rate with a guaranteed mininimum; every job had its standard allowed time. We all manipulated the system, which was strictly forbidden and should have made the system worthless as an incentive. Yet my conclusion at the end was that precisely this informal aspect of the rate system made otherwise stupid repetitive jobs interesting and motivating. The challenge of manipulating the rate to our advantage made us into small entrepreneurs. Formal control systems have informal psychological effects; this discovery would later contribute to my choice of the psychology of budgeting as the subject of my doctoral dissertation.

A final conclusion is about the vast difference in life patterns between the skilled and semi-skilled workers at the machine shop and the unskilled laborers at the cigarette factory. The machine shop personnel led a bourgeois kind of life: well organized and purposeful, with clear norms, even if not everybody respected these. The tobacco workers, on the other hand, were an alienated, marginal, virtually anti-social bunch, which also means one should be careful about generalizing about "the Amsterdam worker"—or any group like that—before really knowing it.

This participant observation fed my interest into the impact of jobs on people, started by Kuylaars's book. This impact depends both on the jobs and on the people; and for the latter, their jobs are only one aspect of their lives—an indispensable but not necessarily the most important aspect. Work behavior cannot be fully understood if work is not put in perspective next to other life interests of workers; and, as I would only understand later, life interests are imbedded in a total culture.

During my work period at the machine shop, Maaike and I got married. I had two days of leave, congratulations from many colleagues, and a money gift from the personnel fund handed out by the chairman of the works committee with a speech at lunchtime. Our honeymoon trip was to the Amsterdam Zoo.

Getting a paid job was no problem in those days: there was over-full employment. I got several offers, but I chose Stork Bros., the large engineering firm where I had spent part of my *HTS* internships. They were a famous builder of large marine diesel engines, located in the town of Hengelo in the Twente region close to the German border. This was the region my male ancestors came from; my father had worked with Stork before he became a teacher. The attractiveness to me of Stork's offer was that the company would first, at their cost, send me for a full year to work with an organization consulting firm, Berenschot, for an all-round training into organization consulting methods; then only I would join Stork itself as an internal consultant.

Berenschot trained me in time-and-motion study; I set piece rates in a small factory of steel furniture, in a cookie bakery, and in the bottling shop of a gin distillery. I worked on a quality control system for a producer of steel bolts and nuts, and on a job evaluation system for the Dutch coffee, tea and tobacco industry. I got insight into their employee selection and training methods. When we moved to Hengelo and I finally joined Stork in September 1956, I had become an all-round junior consultant.

At Stork I lasted for three years (I afterwards found out that my father had also lasted there for three years, a generation before). This was an old and old-fashioned respectable family firm, and a young organization consultant with modern ideas was not warmly received. I remember one occasion when the manager of one of the design departments rushed out of his cubicle to throw me out when he saw me entering his unit.

One of the Stork brothers allowed me an audience about my career prospects, and the next thing I knew was that I was sent on an eight-week General Management course ran by the Dutch Employers Association (*CSWV*). Management courses were just taking off in Holland; it was only the third year that the course I attended was given. This was another great experience: I learned about control, finance, marketing, strategy, industrial relations. The entire program was built around one hypothetical case of a nearly-defunct firm in the hosiery (underwear) business. What I liked most was my classmates: we were sixteen young men from very different businesses, from insurance to brewery, and we learned to know each other very well: our histories, our strengths and weaknesses, our jobs, and our salaries. Which means we became aware of the labor market and our position in it. When I returned to Stork and fell into the same frustration as before, I started reading the employment offered section in the newspaper. I was gone in a matter of months.

The new job I got was Department Manager, groomed to be Plant Manager, of a local hosiery firm: this one made ladies' stockings. Without the management course I would not have considered moving into such a business—but here I was, with the line responsibility for which I would have had to wait another ten years at Stork. Once more I learned—the hard way. There were two owners involved in a constant fight, and I had been hired by one of them to eventually take over from a stout lady, a former factory girl, who had become—as I only found out later—the mistress of the other. The problem was that the lady in question still ran the departments before and after the one I had been made responsible for, and held all the trumps for showing I couldn't do the job. I had naively ran into a trap. I contributed my own blunders as well. I made the classic error of the bright professional in his first management job: trying to do too much myself. I estranged a competent time-and-motion analyst by designing a new standard system much too fast, and I estranged forty floor girls by a lack of patience in listening to them. Within a year I was fired, together with the stout lady; I had served as a pawn so the two owners could make a horse deal and each sacrifice a supporter.

Seeing what was coming, I had already looked around for other jobs. I was hired immediately by N.J. Menko, an integrated textile firm (spinning, weaving and finishing) in the old textile town of Enschede, also in Twente. This was another family-owned company that tried to modernize. They had just extended their top management team of two family members with two outsiders: a successful supersalesman for Marketing, and a professor of economics, Fred L. Polak, for Finance and Manufacturing. Polak felt more competent in the first field than in the second, and he hired me as his assistant for Manufacturing. My new boss was a maverick in Dutch academia; at Rotterdam University he had established himself as a Futurologist, specializing in long-term forecasts; a field to which he eventually returned, after his interlude in the textile industry.

Our growing family—Maaike and I had two sons in the meantime—moved to Enschede. Maaike taught some French classes in a secondary school. We met other young families and made new friends. Working for Fred Polak was a big improvement for me. For the first time I had a boss who backed me up—and of course I made fewer mistakes this time. I became Interim Manager of the weaving mill, and later on General Manager of Staff Services, responsible for construction, maintenance, transport, and methods and procedures.

Polak had brought in a Canadian consulting firm, specializing in textiles, for modernizing the manufacturing organization, and I was their internal contact person. The consultants were competent, hard working people; so hard working, in fact, that one of them died from a heart attack in his apartment while working for us. Their main method was standard setting, and from the experience in working with them I got the inspiration for my doctoral thesis subject some years later.

It had not so much been the firm's name that attracted me to Menko as Fred Polak's reputation as an academic. During the dramatic year at the stocking factory I had slowly made up my mind that I did not want to remain an engineer or a technical manager all my life, but that I should go back to university and get some kind of qualification in the social and economic side of organizations. I had mentioned this to Fred Polak in the hiring interview, and he made me a promise with a Biblical precedent. I refer to *Genesis* 29, where Laban promises his nephew Jacob that if he will work for him for seven years, he will get his cousin Rachel for a wife. Polak promised me that if I would help him for three years reorganizing the firm, he would then give me facilities for an extended study. I got a better treatment than Jacob (who after seven years was put off with Leah and had to work another seven years for Rachel). When the three years were over, in 1963, Polak had already left Menko, but he had made sure that his successor kept the promise, and for the years 1964 and 1965 I was given a 50% unpaid leave of absence.

BECOMING SOMEBODY ELSE: 1963-1980

Some time in 1963 I walked with our eldest son, Gert Jan, who was then six or seven, and I asked him what he wanted to be when he was big. After some reflection, Gert Jan said: "I want to be a scholar (*geleerde* in Dutch)." "Gosh Gert Jan, that is what I would want too!" Gert Jan smiled: "Yes, but in your case, of course, it's too late now!" As a matter of fact, it wasn't.

Up till 1963 the Dutch university system granted doctorates only to persons having a master's level degree in the same field. Thus, with my master's level degree in Engineering I would only be able to get a doctorate in Engineering. For any other doctorate I would have to complete another master's study first, which was almost prohibitive. Fortunately a change of the law in 1963 allowed departments to accept doctoral candidates from any field, subject only to the judgment of the faculty; which opened the way for me towards a doctorate in social science. Doctorates in the Netherlands demanded the completion of a thesis under the direction of a Thesis Advisor (*Promotor* in Dutch), but no further courses or examinations. For a thesis subject I had started with some vague ideas about research into employee motivation. Gradually, remembering my experiences with the incentive scheme in the printing press factory, and later with standard setting in the textile plant, I narrowed my subject down to the psychological impact of standard setting and budgeting on the people involved. After all, these techniques were intended as (among other things) tools for motivation.

When I had come that far, I talked to Fred Polak who suggested that for a Thesis Advisor I should try Herman A. Hutte, a professor of social psychology at the University of Groningen. By sheer accident, I already knew Hutte; I had met him on a a sailing vacation in Friesland (northern Netherlands) during the war, when he was still a student. After testing out my commitment, Herman Hutte accepted me as a doctoral candidate and my subject as a thesis theme. He gave me a list of books to read to fill the gaps in my professional knowledge. One book in particular influenced me greatly: a Dutch textbook on social science research methodology by Professor A.D. de Groot[5]. De Groot was both a mathematician and a psychologist; he wrote a famous doctoral dissertation on the thinking process of chessplayers, which inspired the work of, among others, Herbert Simon in the United States. De

Groot's methodology book showed me the similarities and the differences between the exact sciences, in which I had been trained so far, and the social sciences which I now entered.

With the unpaid 50% leave of absence from my employer I still needed to supplement my income. The family was growing: our third son was born in 1964. I applied for a stipend from the Dutch Productivity Committee (*COP*), that managed the remaining Marshall Aid countervalue funds (the money Dutch industries had paid for goods received under the U.S. Marshall Plan after the war). I got 25,000 guilders divided over two years, which was sufficient to survive. Also, the *COP* secretary wrote a letter of recommendation that was very useful in getting access to companies for my research project.

Menko was not a suitable place to do research in, as I would not be perceived as an impartial observer. With the *COP* introduction I was admitted into six manufacturing plants of five Dutch companies: **Philips Lamp** with two manufacturing units of TV tubes; **AKZO Chemicals** with one manufacturing unit of industrial rayon yarns; **Van Nelle**, a coffee roaster and tea packer; **Samsom**, a large printing and publishing house, and the typewriter manufacturing plant of **IBM** at Amsterdam.

In early 1964 I made a four week trip to the United States, my first. For Menko I visited modern textile plants in New England, the Carolinas and Georgia. Then, under my new 50% leave scheme, I set off to academia and visited the Institute for Social Research in Ann Arbor, Michigan; Cornell University in Ithaca, New York; Yale University in New Haven, Connecticut; and MIT at Cambridge, Massachusetts. I met many of the academics whose work I had read before; for my thesis project two encounters were particularly helpful: with Professor Chris Argyris at Yale, and with Dr. Andrew C. Stedry at MIT. Argyris gave me a copy of his earlier psychological study of budgeting,[6] as far as I know the first ever; and with Stedry I discussed his award-winning thesis on "Budget Control and Cost Behavior"[7] from which I borrowed the key concept of "aspiration levels" as the link between the formal budget and the informal motivation.

After my return home I worked under high pressure, half time for my employer and the other half, plus vacations, for my thesis. I spent over 400 hours in the six plants in semi-structured interviews with line managers of all levels and with staff employees in charge of standard setting and budget administration. One night in November, 1964 I was woken up by a call from Menko: there was a big fire. The plant fire fighting brigade reported to me. It was particularly bad: in one dramatic hour, half of the Finishing Department was gone. I spent the rest of my time at the plant negotiating insurance payments, and supervising the reconstruction.

My research at the IBM typewriter plant led to an invitation for a guest lecture at the IBM World Trade Executive Development Department which at that time was located in Blaricum near Amsterdam. After that lecture, IBM offered me a choice of three jobs, two within the Dutch subsidiary and the third within the international organization; this is the one I chose. I was going to be a trainer in the Executive Development Department, and at the same time I would set up a Personnel Research Department for IBM Europe, under the guidance of the European Personnel Director in Paris and the international Personnel Research Manager in New York. I would join IBM formally in September, 1965, but they would give me four months off for finishing my doctorate thesis.

So I quit Menko, and my life as an engineer. I still hold good memories of this last technical job and of many of my colleagues there, but this was the right time to leave the Dutch textile industry which more and more lost out in international competition. Textile, one of our friends said, is a third-rate business. IBM at that time was definitely first-rate. The contrast could not be stronger.

As a family we moved to Blaricum, to a very nice, bigger house. Our fourth and youngest son was born there in 1967. Maaike found another secondary school, to teach French on a part-time basis.

My doctoral thesis "The Game of Budget Control" was finished in the summer vacation of 1966, and the public defence at the University of Groningen took place in March 1967. I owe a lot to my *Promotor* Herman Hutte. He built up my self-confidence; after all I entered the field of psychology as an amateur; but he took me seriously, and taught me to think as a social scientist. The conclusion of my research was that budget systems potentially held a strong motivating power, because they affected the budgetees' performance aspiration levels. They were not, as some authors suggested, something basically negative from a psychological point of view. The motivation by budget systems was optimal if higher management maintained a game spirit around the achievement of budgetary targets by lower-level managers. Too much pressure led to evasive behavior; not enough attention led to neglect. Participation in budget-setting, then an ideologically popular theme, had some influence on motivation, although less than the literature claimed. For participation, a distinction should be made between financial standards and technical standards. Lower level managers nearly always wanted to have a say in technical standards, but rarely in financial standards. Regardless of whether they had desired to participate, however, if involved they were somewhat more motivated to perform efficiently. The idea of budgeting as a "game" was, in fact, a cultural interpretation, although at the time I had not recognized it as such.

The thesis had been written in English; I saw myself as communicating with the international community, not only with the tiny number of potential interested readers in the Dutch language area. One of my new American colleagues at IBM checked my English. The thesis was well received. I got a *cum laude* qualification, a national award, and two publishing houses, one in the Netherlands and one in Britain, distributed commercial editions of the thesis. Much later I discovered I had become one of the Founding Fathers of a new field called Behavioral Accounting; although I myself deserted this field quickly after siring my intellectual offspring.

My entry in IBM meant a completely new life and interest area. For one thing, my working language now shifted from Dutch to English. The Executive Development Department ran several four-week International Management Schools a year, each for some twenty IBM managers from all over the world. There was an international team of some eight trainers, a group of very skilled secretaries, and a household staff for the residence. I was scheduled as a staff member for some of the schools. The program was a combination of IBM-specific information provided by top managers from the Paris staff, regular business school themes like Finance and Organizational Behavior, and special topics presented by distinguished outsiders, such as an Economist of the European Commission, and an international Trade Unionist.

My special task, as mentioned, was setting up a Personnel Research Department for IBM Europe. I had my own secretary and fairly soon an assistant, a bright young Dutch psychologist. I got my instructions in particular from my functional boss in the United States, and they usually meant travelling; which I loved. Between 1966 and 1971 I was away from home about one third of the time, visiting the Paris office, all countries of Northern, Central, Western and Southern Europe, part of the Middle East, and occasionally the United States. It must be said that IBM was generous: about once a year Maaike was invited to join me, at company expense; in 1967 she came to the United States with me. Most trips were related to the planning and execution of attitude surveys, and meant that I would be staying in a country for several days, and communicate intensely with many of the locals. I met very many people and used all my language skills.

One of the characteristics of the IBM corporate culture was a strong concern of managers with employee morale. This grew out of the convictions of IBM's founder Thomas J. Watson, Sr. many decades before, and it was made one of the basic beliefs ("respect for the individual") in which newcomers were automatically socialized. In this concern with employee morale, attitude surveys were something natural. Incidental surveys of parts of the organization within certain countries had been done by consultants since the 1950s. Each of these used its own method and survey questions. In 1966/67 I headed up an international team of six researchers (three internal, three consultants) who prepared the first internationally standardized questionnaire, for the personnel of IBM's Development Laboratories in six European countries. The questionnaire consisted of 180 standardized items, chosen on the basis of open-ended pilot interviews in the six laboratories, and of a selection from the questionnaires used in earlier incidental surveys. It was issued in five languages and administered in June, 1967. Strong support came from the new Personnel

Research Manager at the IBM World Trade Head Office in New York. He was Dr. David Sirota, an eminent professional as well as a powerful organizer. He had a vision of standardizing and internationalizing surveys throughout IBM World Trade, to create a comparability of results, over time, across countries, and across functions. In November 1967 Sirota surveyed IBM World Trade's total personnel outside Europe and the United States: he covered 26 Asian, Latin-American, and Pacific countries, most of which had not been surveyed before. His questionnaire included many items from our laboratory survey; Sirota pre-tested it on both a U.S. and a Latin American sample.

These two pioneer international projects were followed by surveys of all European and Middle East marketing and administrative operations in 1968 and 1969 and of all manufacturing plants in 1970. Within Europe I had to sell the international survey idea to the countries' functional and general managers: national subsidiaries could not be told to participate. Some top managers were known to have a spontaneous interest in applications of social science and they were approached first; some usually went along with projects that seemed to have the approval of the Head Office and they were approached second; some were expected to be difficult and they were approached last with the strong argument that all the other countries had decided to join. In a few cases this did not work, and the first survey round showed some blank spots for parts of country organizations that could not be surveyed, like in Holland where the General Manager disliked having to listen to another Dutchman. The second time we surveyed, the idea had become so well accepted that almost everyone participated.

IBM USA was not included in these surveys. Its different divisions had survey histories of their own and their own independent survey staffs. Through negotiation with one of his fellow U.S. researchers, David Sirota managed to have the standardized international questions also administered to a sample of U.S. IBM personnel, so that comparisons with the United States could be made.

The international attitude surveys were not sold to IBM managers as a research project, but as a management tool for organization development. Considerable attention was given to data feedback to managers of different levels, as well as to the employees themselves. It was important to avoid defensive reactions of managers whose departments showed unfavorable results. The reactions of nonmanagerial employees, after some initial suspicion, were invariably positive, although sometimes skeptical as to the likelihood of corrective action by their managers. We did several studies about factors that helped and hindered the use of survey data by managers.

In 1969 both David Sirota and I received an IBM Outstanding Contribution Award for our work on the survey program. It was the first time the award went to someone in the Personnel function. The international surveys had become an established practice in IBM. Personnel departments of all major parts of the organization had appointed personnel research officers. Within this new personnel research fraternity we planned a survey approach for the next round, settling for a core set of 60 questions to be used in all surveys, and additional functional and national questions according to need. The core items were selected on the basis of an extensive study of the available literature, and of factor analyses of the results obtained in the first survey round. At the end of the second round, in 1973, some 117,000 questionnaires had been collected from all functions and levels in the company in 72 countries, using 20 different languages.

In the fall of 1970 I had just returned from Beirut on a Friday, and I would be leaving again for the United States on Sunday. On Saturday I drove in the car with Gert Jan, who was then fourteen. He said: "Father, what about this travelling of yours? How long will this go on?" And when I said "How do you mean?" he went on "You see, we don't miss you anymore!" This was strong feedback, and it kept ringing in my ears. My first stop in the United States was Boston, where Professor Robert N. Anthony had invited me for a guest lecture at the Harvard Business School—still about my budget study. After the lecture he asked whether I would be interested in being a guest professor at IMEDE, a business school in Lausanne in the French speaking part of Switzerland, a program for which he did some of the talent scouting. I discussed this later at IBM World Trade Head Office in New York, and it turned out that IBM had a sabbatical policy for which I was eligible. I could even get a leave of absence for two years. Coming home after this trip I told Gert Jan that I had the answer to his question: we would soon travel all together.

In the summer of 1971 we moved to Switzerland, renting a wonderful house in the vineyards overlooking the Lake of Geneva. The boys, then 14, 12, 7 and 4, shifted to French speaking local schools. Gideon, the youngest, played silently in his kindergarten class from the end of August till the end of November. Then he suddenly opened his mouth and spoke French. Maaike's job was the biggest casualty of our move: there was no work for a foreigner to teach the locals French. She remained unemployed for the time we lived in Switzerland. We used all the school vacations for camping expeditions with the whole family, visiting France, Switzerland, southern Italy and southern Spain.

At IMEDE (today called IMD) I learned how to teach a full Organizational Behavior class to international groups of managers in a 19-week course, clumsily at first, more effectively later. I also had time to do research, and with the help of a student assistant from the Technical University I started to do statistical analyses of the IBM attitude survey data base; trying to solve all the puzzles that had come to my mind earlier, but for which the IBM job did not leave me the time. I had recognized that the answers to some of the IBM questions differed from country to country, following a pattern that seemed to repeat itself from one employee category to the next. These were in particular the questions dealing with basic values of the respondents, rather than with their attitudes about daily events in the company. In my IMEDE classes I administered a number of the same questions; and the answers by my management students, none of whom were from IBM, showed significantly similar country differences as I had found for the IBM employees. This proved something that I had suspected all along: the vast IBM employee survey material contained information not only about IBM, but about differences in attitudes between nationals of different countries, IBMers or not. The IBM population was a very specific sample from the national populations, but it was extremely well matched from country to country: same corporate culture, same kind of education level, same kind of jobs, similar age and gender distribution. Serendipitously, the IBM data bank had become a major source of comparative information about the national societies these IBMers came from.

In the summer of 1973 I was due back at IBM. My boss, the European Personnel Director, wanted me to move to the Paris office now, again as Manager of Personnel Research, which had been my title since 1968. I offered a counter proposal. Citing from the results of my recent analysis of the IBM data bank, I proposed to step down to a researcher position, and to do a special study on the meaning of all this for IBM management. My boss, however, said that he needed me in my manager role, and that the project I proposed was too academic to be carried out inside the company; I could give it to some university. Another issue was my date of return. We had planned the realization of an old dream: a three months' expedition of the whole family to Greece and Turkey, leaving on the first of July (when the Swiss schools had closed) and returning on the last day of September (a bit late for the French schools, but they would still accept our sons). My American boss had no sympathy with long vacations and wanted me back in August. This was the last straw; we agreed that I would quit IBM and find a university base to continue my research on the IBM data bank.

I was offered a position at INSEAD, the international business school in Fontainebleau, France, but I hesitated, knowing that INSEAD was a teaching rather than a research place. Then I got another offer from the European Institute for Advanced Studies in Management (EIASM) in Brussels, Belgium, which had been founded in 1971 as an international think tank. This suited my research objectives, but I did not want to quit teaching altogether. I ended up with a two-year contract for two half-time jobs, one teaching at INSEAD and one researching at EIASM. We would move our home to Brussels.

First, however, there was the Greek-Turkish Odyssey. We bought a Ford Transit van, loaded it with our camping equipment, stored our furniture in Lausanne and crossed the Alps. We crossed from Brindisi, Italy, to Corfu, Greece, all through Greece, over to Crete, back to Athens, around the Egean

Sea to Turkey, and all the way through Anatolia to our most distant point, Mount Nemrut. We returned though Bulgaria and Yugoslavia, via Sarajevo and Mostar, which at that time were lively and beautiful. After initial struggles we developed a smooth group dynamics among the six of us, in which the simple tasks of camping life were quickly and effectively performed. In the countries we visited, mothers of sons have a high status, and Maaike enjoyed hers as the mother of four. We had many exciting vacations before and after, but this trip was the ultimate adventure. Each of the boys remembers it as an existential experience.

In Brussels we rented a large town house. Gert Jan left us; he had decided that he wanted to finish his secondary school in Holland, staying with relatives. The other three sons continued their education in French at local schools. Once more Maaike could not teach French, but in Turkey she had become interested in the Islamic world, and at the nearby Free University of Brussels she enrolled in Islamic Studies: Arabic, Turkish and Persian language, history and culture.

At INSEAD I became a staff member of CEDEP (*Centre d'Éducation Permanente*), providing post-experience training to managers from an international consortium of companies. Most of these, however, were French multinationals, and I was asked to teach part of my classes in French. I bribed a kind participant, promising him a drink at the bar every night if he kept a list of my recurrent language errors, and I think we succeeded in weeding out the worst errors this way. I enjoyed teaching to high-calibre experienced participants; I also enjoyed an eminent group of colleagues who stimulated my thinking, in particular André Laurent, M. Sami Kassem and Owen James Stevens.

In Brussels, at EIASM, I found the time and the quiet to really dig into the IBM data. There was a good library where I hunted for related literature; I had computer support from a service bureau, and access to a terminal where I could do the simpler analyses myself in timesharing. I exchanged views with colleagues from other European countries, and ventilated the progress of my own work through the Institute's Working Paper series—which was a way of getting feedback from the selected circle of colleagues who read them. I attended workshops, and served on the thesis committees of doctoral candidates in different countries.

My initial estimate had been that the two years of the contract would be enough to write up the conclusions from the IBM study. However, the job appeared much bigger than I had realized. I made fundamental discoveries and had to make fundamental decisions about the conduct of the study. One discovery was that I was no longer dealing with psychological issues; what I studied were manifestations of the national cultures of countries. **Culture** became a key word in my study: I was practicing the comparative anthropology of modern societies. The relevance of culture for organizations and management was not as obvious in the 1970s as it became later. Using a metaphor from the then increasingly popular computer field, I coined my own household definition of what I meant by "culture:" **the collective programming of the mind which distinguishes the members of one group or category of people from another**.

The difference between a psychological and an anthropological approach to the data became operationally clear in the choice of the level of analysis. As a psychologist I was accustomed to correlating the answers of individual respondents, that is to analyse at the level of individuals. In comparing countries it made more sense to correlate mean scores for the respondents from these countries. This is not a trivial matter: the two approaches can lead to entirely different results. Other categories for aggregating the data, like occupation, again lead to different correlations, and therefore different conclusions. Correlating individual data appeals to psychologists, correlating categories to sociologists, correlating country data to political scientists and anthropologists. As a former engineer I had never been a complete psychologist, which may have made it easier for me than it would have been for some other psychologists to switch to the level of analysis of another social science discipline.

Moving from an individual-level to a country-level analysis did simplify my task in one way: I was no longer dealing with 117,000 individual responses, but with responses from 40 countries (from the altogether 72 countries for which I had some data, I initially took only the ones with more than fifty IBM respondents per survey round per country, to be statistically conservative).

It was obvious that I would have to extend my research time beyond the two years of my contract with EIASM, and I applied for financial support from IBM. Not from the Personnel Department this time, but from the University Relations office. IBM granted US$ 50,000 to support my research. For the next four years, from 1975 through 1979, I spent most of my time researching, reducing the teaching at CEDEP/INSEAD to about 20%.

In the meantime at INSEAD I had been involved in a side project. My colleague Sami Kassem, visiting professor from the United States, proposed to organize a meeting of senior European scholars in the field of organization studies. Kassem was an Egyptian by birth and initial education, and he found the European developments more relevant to his cross-cultural experiences than what he had learned in America. We got CEDEP to sponsor a European workshop on Organization Theory in May 1975. Contributors were the leading scholars in the field of those days, like Michel Crozier from France, Renate Mayntz from Germany, and Derek S. Pugh from Britain, and the purpose of the workshop was to present a state of the art. The proceedings appeared in a volume *European Contributions to Organization Theory*, edited by Kassem and me, in the beginning of 1976.

It was around 1975 that I started to think of the country differences in the IBM data bank in terms of **dimensions**. I had first analyzed the answers on questions related to the distribution of power. This led to the identification of a characteristic of national societies that I labelled Power Distance, ranking from relatively equal (like Denmark) to very unequal (like the Philippines). Answers to other questions in the IBM data base also showed stable distribution patterns across countries, but not the same pattern as Power Distance. Searching further I identified three more dimensions: Uncertainty Avoidance, Individualism-Collectivism and Masculinity-Femininity. Across the 40 countries, the last three were uncorrelated (orthogonal). Individualism was negatively correlated with Power Distance; but I hesitated to consider them as aspects of the same dimension because the two were conceptually distinct. In fact the reason for the correlation was external: country Individualism scores were positively correlated with national wealth, and Power Distance scores negatively; and if I kept wealth constant (comparing rich with rich and poor with poor countries), the correlation between Power Distance and Collectivism disappeared.

Finding significant correlations of Individualism and Power Distance scores with wealth (Per Capita Gross National Product) meant external validation: it showed unambiguously that the responses of the IBM employees contained information about their countries, not specific to IBM. I started searching the literature for other quantitative results of country comparisons that could be expected to correlate with the IBM country dimension scores. I had opened a Pandora's box: there was much more related literature than I had ever dreamt of, in sociology, anthropology, psychology, political science, market research, demography, economics, even comparative medicine. In the end I had found some forty studies external to IBM that showed country scores significantly and meaningfully correlated with the IBM dimensions. The pattern of external validations, of course, helped considerably in interpreting the dimensions.

The theoretical meaning of the dimensions is, I believe, that they reflect basic problems that every human society encounters, and for each of which a range of solutions is possible. Power Distance stands for social inequality, Individualism for the relationship between the individual and the collective, Masculinity for the social role distribution between the sexes, and Uncertainty Avoidance for coping with the unknown future.

My choice of the label Masculinity-Femininity has been criticized a lot, funny enough almost exclusively in countries which I found to score "masculine." I think I had good reasons for my choice. This dimension was the only one for which the country scores depended strongly on the gender of the respondents. Country scores on Femininity correlate with several indexes of gender role distribution in societies, such as the percentage of women in elected political offices, women in higher professional roles, and the extent to which women and men pursue the same degrees in higher education. The resistance to the term showed that I hit upon a culturally sensitive issue; fine, what else was to be expected?

Altogether, the analysis of the IBM data bank during my six years at the Brussels EIASM filled seventeen working papers. At the end, I re-worked these into a book manuscript "Dimensions of National Cultures," and wrote to nineteen American publishers asking for their interest. Only half of them reacted, mostly by a letter in the sense of "your manuscript looks very interesting but it does not fit our list." One asked to see the manuscript. In the meantime, however, I had attended the fourth conference of the International Association for Cross-Cultural Psychology, of which I had become a member. The conference was held at Munich, Germany, in the summer of 1978. There I met Professor Harry C. Triandis from the United States whom I told about my project. Harry, an eminent and broad scholar, immediately recognized the potential importance of what I had to offer. He spoke to Walter J. Lonner and John W. Berry who had been appointed Series Editors for a new series with Sage Publications, Beverly Hills, California, on "Cross-Cultural Research and Methodology." Sage was also one of the nineteen publishers I had written to myself. By the same mail that I received their "... it does not fit our list" letter from a lower authority, there was another letter from the Publisher and President, Sara Miller McCune, expressing her interest in publishing the book, at the recommendation of Lonner and Berry.

Sara and I met in London in 1979. We spent hours in an Italian restaurant discussing the project. One problem was that the manuscript was far too long for a regular social science publication, but Sara showed herself flexible: she believed in the book. We arrived at a compromise that I would leave out some appendixes and Sara would use a smaller letter and a larger number of pages than normal (475, as it turned out). Also, Sara suggested another title: *Culture's Consequences*. The book appeared in December 1980.

In 1979 I spent most of the spring at the International Institute for Applied Systems Analysis (IIASA) at Laxenburg Castle near Vienna, Austria. The family stayed in Brussels. IIASA was an East-West institute, hosting scholars from either side of the Iron Curtain that in those days still separated the communist from the capitalist world. Working and partying side-by-side with Hungarians, Poles, Czechs, East Germans and Russians opened new perspectives.

In many respects, 1979 was a breaking point. My book had gone to the printer, my contracts with CEDEP and EIASM would once more expire, the rent contract for our house in Brussels ran out, and the family decided that after eight years they wanted to go home to Holland. There was only one small problem: I was fifty-one and I did not have a job. I tried for a professorship at one of the Dutch universities but with cuts in the education budget, departments were not very interested in a compatriot returning from abroad. Yet at the end of the summer we moved, and rented a newly built house at Zwijndrecht, across the river from Dordrecht, overlooking one of the busiest waterways of Europe. Our main reason for choosing this place was that Maaike had attended the *Gymnasium* at Dordrecht, and it seemed to be still a good school for our two youngest sons Bart and Gideon. The second son, Rokus, had completed his *baccalauréat* (secondary school certificate) in Belgium and now studied at the University of Groningen in the Netherlands.

For a living I made some money from guest teaching, among others at INSEAD, and from a research contract. Maaike, who in Brussels had graduated from her Islamic studies with a master's level degree, got back into teaching French at two secondary schools for adult education. One day in the fall of 1979 a headhunter approached me with a job offer. Here I was back in the business world: at the beginning of 1980, I found myself Director of Human Resources of Fasson Europe at Leiden, Holland.

MAKING SENSE: AFTER 1980

Fasson was a division of Avery Label, a U.S. manufacturer of self-adhesive products. Fasson produced the material from which labels are cut, in large rolls on huge machines resembling those used for making paper. Fasson Europe handled all Avery's non-American base material business, in fourteen European countries plus Australia, Brazil and South Africa.

My area of responsibility as a Director of Human Resources was all of Fasson's personnel in these seventeen countries. This was a new function: my predecessor had only been responsible for the Dutch company which employed about half of Fasson's 1,700 employees. The unique thing about my employment contract was that I would be part-time: I had negotiated that I would work for Fasson three days a week, leaving the remaining time free for my research and teaching obligations. I had been surprised that they accepted. What helped was the novelty of the job; nobody really knew how much of a job it would be. My boss, the Group Vice President, was a culturally sensitive American, prepared to try unorthodox solutions abroad.

In the Fasson job I liked being part of a top-management team, getting insight into all aspects of the business: not only the personnel but also the strategy, the product, the technology, the research, the markets, the competition, the suppliers, the finance, the control systems. Fasson was a well-run company, but relatively small. In our organizational roles we were not as specialized as, for example, IBMers. Being a big frog in a small pool suited me better than being a small frog in a big one. I travelled to all the European subsidiaries but left the visits to the overseas subsidiaries to the line managers; we were cost conscious about travel. It was fascinating to see the differences in functioning from one subsidiary to another. For example, our manufacturing plants in Belgium, Britain, France and Holland all had some form of joint consultation between management and employees. I sat in with all four, but the dynamics of the meetings were vastly different, nicely following the Power Distance distinction I had identified in my research: active involvement of the employees in the discussion in Britain and Holland, passivity in Belgium and France.

The Fasson episode lasted until 1983. I could not very well see myself as a Human Research Director until retirement. When at Fasson my boss was promoted back to the United States, his successor wanted me to work full-time, which did not suit me. When I got an offer for a more academic job elsewhere, I accepted.

The new job was Dean of the Senior Management Forum (Semafor) at Arnhem. Semafor (it was not called that when I took over, I coined the new name) was a nonprofit foundation, in existence since 1953, that ran five-week residential management courses for senior managers, as well as short conferences for outside Directors (*commissarissen*). Except for a Dean, an administrator, and two secretaries, Semafor had no staff of its own; we used faculty from Dutch universities, and distinguished speakers from companies and public bodies; conferences were held at a countryside hotel. Semafor had a good reputation and a captive clientele from medium-size business companies and public organizations, including the Dutch army and airforce.

We bought a nice bungalow near a forest at Velp, a suburb of Arnhem, and moved there in 1984. With some difficulty, Maaike could continue her teaching from Velp; our son Bart had by then finished his secondary school, and Gideon stayed with a schoolmate in Dordrecht to finish his.

The Semafor job brought me back into the academic world, and at the same time in contact with the Dutch business world. I made a number of innovations in the program which were necessary, as it had become somewhat dusty and was losing customers to courses abroad. My contract was for four days a week, again leaving some time for my scientific work. However, after two years the work had become routine, and when I was finally offered a professor's chair, at the University of Limburg at Maastricht in 1986, I did not hesitate.

Going back to 1980 for a moment, with my two days a week off from Fasson I wanted to continue to do research, but I needed an institutional umbrella for that. Then I met Bob Waisfisz. Bob was a Dutchman with a university education in business administration and anthropology, who had worked in Lebanon and Turkey for the International Labor Office, and later on had run a Dutch non-governmental organization for Development assistance. Bob recognized the importance of my work for intercultural communication even before my book had appeared. In fact I think he believed more in it than I did myself. Bob suggested we could create our own research institute, as a nonprofit foundation. On December 16, 1980, the Institute for Research on Intercultural Cooperation (IRIC) was founded. Bob was Chairman, and I was Director. Our secretariat was at an institute in Delft where Bob worked at the time and where we paid his secretary for the hours she needed for IRIC.

IRIC was definitely nonprofit. During the first years we were glad when we could pay the out-of-pocket expenses, and our own time was unpaid. When I moved from Fasson to Semafor in 1983, the IRIC secretariat moved with me, to a small cubicle in the Semafor office. We engaged an unemployed teacher as a volunteer secretary. In 1984 the business had become so much better that we could pay her a salary. It was not until 1988 that IRIC would become a going concern, able to pay even its Director.

In late December, 1980 I attended the fifth conference of the International Association for Cross-Cultural Psychology, which this time was held at Bhubaneswar, India. It was one of the nicest conferences I ever participated in, held on the campus of a local university where students and holy cows shared the lawns. I found myself assigned as a discussant to a session in which a Dr. Michael H. Bond from Hong Kong presented a joint paper on "Human Values in Nine Countries," based on students' answers to a U.S. questionnaire, the Rokeach Value Survey. Looking at the paper I found it to contain a basic level-of-analysis flaw: countries had been compared on indexes designed for comparing individuals. In "Culture's Consequences" I had called this the **reverse ecological fallacy**. In the discussion I argued that the authors had committed this fallacy, although I was afraid that the presenter would not love me for this criticism.

During the session there had been no time for a rejoinder, but in the evening there was an open air reception for the conference participants. In the darkness a tall figure emerged behind me; it was Michael Bond. "I have reflected on your comments of this afternoon," he said. "I think you were right." This was the historic beginning of an intellectual partnership that lasted until today. Michael did a re-analysis at the country level of the data from his presentation, and found them to reflect all four dimensions from my work. We reported this in a joint article published in 1984.

In the summer of 1981 I made my first visit to Japan, as a guest of the Japan Foundation, to attend a conference on the internationalization of Japan. The conference and my trip had been organized by Professor Hiroshi Mannari of Kwansei Gakuin University in Nishinomiya, who later became the chief translator of *Culture's Consequences* into Japanese. From Japan I flew to Beijing, China, on an invitation by the recently created Chinese Enterprise Management Association (CEMA). The liberalization of the Chinese economy had only started in 1978. An English friend, Bob Garratt, had made the Chinese aware of the importance of a culturally relativist look at management. I lectured to CEMA in Beijing and Shanghai and also visited the Guangzhou chapter in the South.

From Guangzhou I travelled to Hong Kong by hydrofoil; Michael Bond met me at the quay, and offered hospitality at the Chinese University. Our discussions dealt with cultural relativity and with the way our minds are pre-programmed by culture, including the minds of the researchers themselves. In those days Michael conceived his Chinese Value Survey project: recognizing that the results of surveys designed by Westerners were biased by their Western minds, he asked his Chinese colleagues to compose a questionnaire with a deliberate Chinese mental bias, and administered this around the world in Asian and non-Asian countries. I helped Michael in finding a translator from Chinese to Dutch, and in motivating a colleague to administer the questionnaire to a sample of students in Holland. The Chinese Value Survey produced also four country-level dimensions.[8] Three were significantly correlated with three of the IBM dimensions, but no counterpart of Uncertainty Avoidance was found using the Chinese questions; instead, there was a new dimension which Michael Bond called "Confucian Dynamism" and that I later re-baptized Long-Term Orientation. Long-Term Orientation as a values complex correlated with national economic growth in the past 25 years; as far as we knew, this was the first values measure ever that did. This was an exciting finding; it became the theme of our next joint publications (1988, 1989, 1991).

IRIC also started to produce interesting research. In 1981-82 I did a study for a Dutch management consulting firm that had started an office in Jakarta, Indonesia. Indonesia did not figure among the forty countries from the IBM data base that I had studied for *Culture's Consequences*. There was an Indonesian sample in the IBM data, but I had initially omitted it, because it was (just) below the limit I had set of fifty respondents per survey round. I decided to lower the limit to twenty, which meant I could get scores on the four dimensions for another ten countries, plus for three multi-country regions assumed to be culturally more or less homogeneous: Arabic-speaking countries, East Africa, and West Africa. The country scores obtained in this way aligned themselves perfectly with the scores of the other countries, both impressionistically and in their correlations with outside indexes. From this moment onwards, I used the scores for 53 (50 + 3), rather than for 40 countries in my writings. The newly obtained score for Indonesia served as the basis for an in-depth comparison with the Netherlands. I revisited Indonesia for the first time since 1947, interviewing a number of Indonesians in Jakarta. The result was a booklet on *Cultural Pitfalls for Dutch Expatriates in Indonesia*, issued to its customers by the consultant, Twijnstra Gudde International. According to Indonesian readers, it was equally useful for making Indonesians aware of cultural pitfalls in dealing with the Dutch; which was exactly what I had hoped for. The text was also published in two articles in 1982.

For IRIC, Bob Waisfisz did a study of the problems encountered by Dutch expatriates in Saudi Arabia. The study was commissioned by a large Dutch contractor, Ballast-Nedam, that did a huge building project for the Saudis. On the basis of his research, Bob designed a training program for future expatriates that used the four IBM dimensions (called the 4-D model) as a basis. When Ballast-Nedam's project ended, the program was opened for participants from other companies. These trainings became quite a big operation, and in 1985 we decided that Bob would found a separate training institute, ITIM (Institute for Training in Intercultural Management), located in The Hague, while I retained the research activities within IRIC in Arnhem. ITIM would pay a percentage of its proceeds to subsidize IRIC's continued research.

Culture's Consequences sold well. Sage and I had the good luck that the theme of national cultural differences, which had been an exotic hobby before, just became popular at the time the book appeared; and there was little serious competition. The reviews were mixed, some ecstatic, some irritated, condescending, or ridiculing. I had really made a paradigm shift in cross-cultural studies, and as Thomas Kuhn[9] has shown, paradigm shifts in any science meet with strong initial resistance. The agreement with Sage was that we would follow the 1980 hardcover integral edition with an abridged paperback version. I had prepared the paperback text for this already in 1981, leaving out a lot of the methodological and statistical proofs and the tables of base data; those who wanted to check my calculations could consult the hardcover edition. Because the hardcover sold so well, Sage postponed the paperback until 1984. As the manuscript dated from 1981, the scores for the 13 additional countries, calculated in 1982, are not included in the paperback, whereas they do figure in other publications since 1983. Paperback and hardcover together were still selling well in the mid-1990s; the book has become a classic.

Another topic that became popular in the early 1980s was organizational, or corporate, culture. Paradoxically, the IBM studies, although based on data from a corporation, did not yield any information about IBM's corporate culture, for lack of comparison. Right after finishing the manuscript for *Culture's Consequences* I had already made a sketch design for an organizational culture study I would like to do: rather than, as in IBM, study one organization in many countries, I would study many organizations in one country, the Netherlands. However, this was logistically much more difficult than the IBM studies. I could not rely on the infrastructure of the corporation, but I would have to gain access to the different organizations one by one. Moreover, these participating organizations would have to pay for the research; IRIC was unsubsidized. I started actively seeking prospects for an organizational culture study after I had joined Semafor in 1983; my new industry contacts gave me an idea where to look for partners. Selling the project was a hard job. Everybody I approached wanted to know whether Philips Lamp would participate. Philips were not only the country's largest corporation, but also considered leading in the field of management methods. Of course I had asked Philips; but their decision process was painfully slow, and in the end they would be the last to join.

One day in 1984 two visitors from a Danish consulting firm walked into my office. They were extremely interested in the organizational culture project, and offered to contact a number of good Danish companies on our behalf. Thus the project came to cover two countries, Holland and Denmark. We ended up with twenty units from ten organizations, varying from two police corps in Holland to the Lego toy plant in Denmark. IRIC leased two researchers from the University of Groningen, Dr. Geert Sanders and Bram Neuijen. In Denmark a friend from the Association for Cross-Cultural Psychology, Denise Daval Ohayv, made herself free to provide the local coordination. We did interviews in all twenty units, employing eighteen graduate students and young professionals as interviewers whom we trained. The next phase was a survey study for which the questionnaire was based partly on the interviews and partly on the IBM cross-national study. The project ran in 1985 and 1986; it led to several publications, of which the most complete appeared in *Administrative Science Quarterly* in 1990. The key conclusion was that organization cultures are not, as the faddish management literature claimed, a matter of shared values among an organization's members. They are a much more superficial phenomenon: shared organizational practices, developed on the basis of the values of founders and significant top leaders.

The University of Limburg at Maastricht where I got my professorship, in the far South-Eastern tip of the country, was the youngest university of the Netherlands, founded only in 1974. The department of Economics and Business Administration started in 1984, and in early 1985 I was invited for a small Visiting Professorship, two days a month, the content of which I could define myself. I called it **Organizational Anthropology** (in Dutch: *Vergelijkende Cultuurstudies van Organisaties*). This was in addition to my jobs at Semafor and IRIC in Arnhem. Then, in early 1986, I was asked to set up a new program in Management, which would be a full job; I argued successfully that this program should focus on International Management. So I left Semafor, and as of mid-1986 I occupied two professorships in the same department, Organizational Anthropology and International Management. I rented a small appartment in Maastricht which is a beautiful old city, and commuted once a week from and to Velp, a three-hour train ride.

On May 15, 1987 I held my inaugural address, in Dutch, with a title that can be translated as "Dutch Culture's Consequences: Health, Law and Economy." Medicine and Health, Law, and Economics (including Business Administration) were the main Departments of the University at that time; I tried to show for the case of the Netherlands that the same cultural forces affect all these fields.

A few weeks before, our first grandchild had been born. More grandchildren were born later. Being a grandparent is a great experience. One has all the fun of seeing children grow up, but not the problems of parenthood. It is like re-living one's own childhood.

IRIC was initially left in Arnhem; I spent one day a week at the IRIC office before or after going to Maastricht. In 1988 we reached an agreement with the University of Limburg that IRIC would be brought under their umbrella, as a financially independent unit. The university would provide housing and secretarial support in exchange for a seat on the Board, and the promise that IRIC would use university faculty and students for its projects.

My professorship gave me the time to write a popular book on culture. *Culture's Consequences* was a scholarly volume, too difficult for students and lay readers; this applied also to the abridged 1984 edition, although some professors tried to use it as a textbook for lack of better. Having trained various publics in intercultural communication, I had some idea of the desired level of sophistication of a new book. It became an update of the message of my earlier research, plus the results of Bond's Chinese Values Survey study, plus IRIC's Organizational Culture study, plus an extended part on applications and implications. I had separate chapters of the manuscript read by non-expert friends and students for clarity. The result was *Cultures and Organizations: Software of the Mind*, published in hardcover by McGraw-Hill UK, London, in 1991, and in paperback by Harper Collins UK in 1994. The book was well received, and up till the end of 1995 had been translated into ten other languages.

In the academic year 1990-91 I was granted a sabbatical. Maaike had just retired as a teacher, so we could travel together. I taught for one semester at the University of Hong Kong, and the second semester at the University of Hawaii at Manoa. In Hong Kong I was made a Honorary Professor, which meant I could serve on thesis committees. In Hawaii I had earlier, in 1985 and 1987, taught summer courses. We made our trip around the world where possible by train, from Holland to China through Siberia and Mongolia and on the way back through Canada. My Maastricht colleague Arndt Sorge in the meantime acted as IRIC's Director.

After my return to Maastricht I took over IRIC again, which was very busy with four different research projects, and I taught at the university for another two years. On October 1, 1993, the day before my 65th birthday, I became an Emeritus. The Department organized two international workshops on the occasion, one in June on "Accounting and Culture," also celebrating my past in Behavioral Accounting, and one at the end of September on "Intercultural Cooperation in Europe," concluded by my Valedictory Address (farewell lecture) "Images of Europe"—this time given in English.

Since my retirement I have been a Visiting Professor at Hong Kong again, in late 1993/early 1994 and in the fall of 1995, and in Hawaii in the fall of 1994. In the Netherlands I became a Fellow of the Center for Economic Studies of Tilburg University. I also act as the Chairman of IRIC, which under its new Director, Dr Niels Noorderhaven, is now affiliated with two universities: Maastricht and Tilburg.

An analysis of citations shows that my work on culture has reached a wide public. Two thirds of the citations of my name in the *Social Science Citation*

Index (based on professional journals only) refer to *Culture's Consequences*, showing that a book is a much more effective way of communicating ideas that any article. From 1980 to the end of 1994, 834 citations (excluding self-citations) of the book have been listed. The variety of the citing journals is surprising. For the five-year period 1989-1993, on a total of 434 citations, 158 different journal titles appeared.

The five journals providing the most citations were in psychology, management, communications and organization sociology: *Journal of Cross-Cultural Psychology, International Journal of Psychology, Journal of International Business Studies, International Journal of Intercultural Relations*, and *Organization Studies*. Together they supplied 29% of all citations.

Fifty journals accounted for 2-9 citations each. Among them were: *Accounting, Organizations and Society; Academy of Management Review; Administrative Science Quarterly; Communication Education; International Journal of Comparative Sociology; International Journal of Public Administration; Journal of Marketing; Journal of Nursing Administration; Sex Roles; Sociologie du Travail* (France); *Strategic Management Journal; Technological Forecasting and Social Change.* Together they accounted for 47% of all citations.

A hundred and three journals carried just one citation; together 24% of all citations. They included: *Annals of Tourism Research; Anthropos; Behavioral Sciences and the Law; Brookings Papers on Economic Activity; Business History; Journal of Peace Research; Dynamische Psychiatrie* (Germany); *Economic Geography; European Journal of Operational Research; Health Education Quarterly; Historisk Tidsskrift* (Norway); *Journal of Genetic Psychology; Journal of Labor Economics; Journal of Nervous and Mental Disease; Language in Society; Public Relations Review; Revue belge de philologie et d'histoire; Scientometrics; South African Journal of Philosophy; Systems Practice.*

Looking at the content of the citations I find that different disciplines tend to cite different dimensions. Psychologists most often refer to Individualism/ Collectivism. Current psychological theories were developed in individualist Western cultures. The assertion that most nonwestern cultures are collectivist and therefore expect people to behave as ingroup members rather than as individuals appeals especially to psychologists from newly industrializing countries. The introduction of the Individualism/ Collectivism dimension has led to a contingency approach to various psychological theories previously assumed to be universal, like Maslow's model of human needs. Besides many articles, two books have been devoted to this dimension alone, by Kim et al. and by Triandis.[10]

Sociologists and management researchers have spread their interest more across all four or five dimensions; if they have a special interest, it is rather in Power Distance and Uncertainty Avoidance. These dimensions explain the different preferences in different countries for centralization and formalization. Development economists have been most interested in Individualism/ Collectivism and Long/Short Term Orientation, associated respectively with wealth and with economic growth.

This is not the story of a planned career; I have been hopping back and forth between practice and theory, and from one discipline to another. It was not very orderly, and I know there are people who dislike that. It was not deliberate; it just happened; but then the world around us also just happens. At least we humans don't know its plans. Looking back I see a messy, hopscotch hike, but it made sense.

PUBLICATIONS

Listed are only first publications of a professional nature. From the publications in other languages than English only a selection has been included. Articles marked * have been assembled in the book *Uncommon Sense about Organizations* (1994).

1964

Arbeidsmotieven van volontairs [Work goals of student trainees]. *Mens en Onderneming, 18*(6), 373-91.

1967

The game of budget control. Assen NL: Van Gorcum; 1968, London: Tavistock Publications [commercial version of Doctoral Dissertation, University of Groningen]. Annual Efficiency Award 1968, Dutch Management Association (*NIVE*).

1968

Baas en Budget [*Manager and budget*]. Alphen aan den Rijn NL: Samsom.

1970

With T.A. Lindner. Messung sozialer Distanz [Measuring social distance]. *Gruppendynamik, 23*(4), 337-56.

1972

With P.J. Sadler. Leadership styles: Preferences and perceptions of employees of an international company in different countries. *Mens en Onderneming, 26*(1), 43-63.
*The colors of collars. *Columbia Journal of World Business, 7*(5), 72-80.

1973

*Frustrations of personnel managers. *Management International Review, 13*(4-5), 127-43.
Employee surveys—A tool for participation. *European Business*, Autumn, 62-69.

1974

With R.Y. Kranenburg. Work goals of migrant workers. *Human Relations, 27*(1), 83-99.
*Predicting managers' career success in an international setting: The validity of ratings by training staff versus training peers. *Management International Review, 15*(1), 43-50.

1975

*Perceptions of others after a T-group. *Journal of Applied Behavioral Science, 11*(3), 367-77.

1976

[Editor] With M.S. Kassem). *European contributions to organization theory.* Assen NL: Van Gorcum.
The importance of being Dutch: National and occupational differences in work goal importance. *International Studies of Management and Organization, 5*(4), 5-28.
*Alienation at the top. *Organizational Dynamics, 4*(3), 44-60.
Nationality and espoused values of managers. *Journal of Applied Psychology, 61*(2), 148-55.
*With L.I. Rajkay. Looking at the boss and looking at ourselves. *Management International Review, 16*(2), 61-71.
*People and techniques in budgeting. In C.B. Tilanus (Ed.), *Quantitative methods in budgeting* (pp.10-22). Leiden NL: Martinus Nijhoff.

1977

[Editor] "Power in organizations," special issue of *International Studies of Management and Organization, 7*(1).

*Confrontation in the cathedral: A case study on power and social change. *International Studies of Management and Organization, 7*(1) 16-32.

*With L.A. de Bettignies. Communauté de Travail "Boimondau." *International Studies of Management and Organization, 7*(1), 91-116.

Cultural elements in the exercise of power. In Y.H. Poortinga (Ed.), *Basic problems in cross-cultural psychology* (pp. 317-328). Amsterdam: Swets & Zeitlinger.

1978

*Businessmen and business school faculty: A comparison of value systems. *Journal of Management Studies, 15*(1), 77-87.

*Occupational determinants of stress and satisfaction. In B. Wilpert & A.R. Negandhi (Eds.), *Work organization research: European and American perspectives* (pp. 233-252). Kent OH: Kent State University Press.

*The poverty of management control philosophy. *Academy of Management Review, 3*(3), 450-61.

Culture and organization—A literature review study. *Journal of Enterprise Management, 1*(1/2), 127-135.

*Amsterdamse arbeiders in 1955: een participerende waarneming [Amsterdam workers in 1955: A participant observation]. *Amsterdams Sociologisch Tijdschrift, 5*(3), 516-39.

1979

[Editor] *Futures for work*. The Hague: Martinus Nijhoff.

Value systems in forty countries: Interpretation, validation and consequences for theory. In L.H. Eckensberger, W.J. Lonner, & Y.H. Poortinga (Eds.), *Cross-cultural contributions to psychology* (pp. 389-407). Lisse, NL: Swets & Zeitlinger.

*Humanization of work: The role of values in a third industrial revolution. In C.L. Cooper & E. Mumford (Eds.), *The quality of working life in eastern and western Europe* (pp. 18-37). London: Associated Business Press.

Hierarchical power distance in forty countries. In C.J. Lammers & D.J. Hickson (Eds.), *Organizations alike and unlike: Towards a comparative sociology of organizations* (pp. 97-119). London: Routledge & Kegan Paul.

1980

Culture's consequences: International differences in work-related values. Beverly Hills, CA: Sage. Abridged paperback edition 1984. [Translations: Japanese, French, Italian (the latter two with D. Bollinger)].

*Angola coffee—Or the confrontation of an organization with changing values in its environment. *Organization Studies, 1*(1), 21-40.

Motivation, leadership, and organization: Do American theories apply abroad? *Organizational Dynamics, 9*(1), 42-63.

1981

Do American theories apply abroad? A reply to Goodstein and Hunt. *Organizational Dynamics, 10*(1), 63-68.

Management control of public and not-for-profit activities. *Accounting, Organizations and Society, 6*(3), 193-221.

1982

Cultural pitfalls for Dutch expatriates in Indonesia. *Euro-Asia Business Review, 1*(1), 37-41; *2*(1), 38-47.

Skandinavisk management i og uden for Skandinavien [Scandinavian management within and outside Scandinavia]. *Harvard Børsen, 2*(Spring), 96-104.

Dimensions of national cultures. In R. Rath, H.S. Asthana, D. Sinha, & J.B.P. Sinha (Eds.), *Diversity and unity in cross-cultural psychology* (pp. 173-187). Lisse, NL: Swets & Zeitlinger.

Cultural dimensions for project management. In J.O. Riis, J. Lauridsen, M. Fangel, S. Hildebrandt, & F. Runge (Eds.), *Project management–Tools and visions* (pp. 683-700). Copenhagen: Danish Technical Press.

Energy and human nature. *Indian Psychologist, 1*(2), 1-9.

1983

Dimensions of national cultures in fifty countries and three regions. In J.B. Deregowski, S. Dziurawiec, & R.C. Annis (Eds.), *Expiscations in cross-cultural psychology* (pp. 335-355). Lisse, NL: Swets & Zeitlinger.

The cultural relativity of organizational practices and theories. *Journal of International Business Studies, 14*(2), 75-89.

Japanese work-related values in a global perspective. In H. Mannari & H. Befu (Eds.), *The challenge of Japan's internationalization: Organization and culture* (pp. 148-169). Nishinomiya: Kwansei Gakuin University and Kodansha International.

Culture and management development. Discussion Paper MAN DEV/28. Geneva: International Labour Office, Management Development Branch.

National cultures revisited. *Behavior Science Research, 18*(4), 285-305.

1984

Cultural dimensions in management and planning. *Asia Pacific Journal of Management, 1*(2), 81-99.

The cultural relativity of the quality of life concept. *Academy of Management Review, 9*(3), 389-98.

With M.H. Bond. Hofstede's culture dimensions: An independent validation using Rokeach's value survey. *Journal of Cross-Cultural Psychology, 15*(4), 417-33.

1985

American and Dutch business values—Similarities and differences. In P. Sanders & J.N. Yanouzas (Eds.), *Managing in a global economy* (pp. 11-14). Tilburg, NL: Tilburg University.

The interaction between national and organizational value systems. *Journal of Management Studies, 22*(4), 347-357.

1986

[Editor] "Organizational Culture and Control," special issue of *Journal of Management Studies, 23*(3).

The role of cultural values in economic development. In L. Arvedson, I. Hägg, M. Lönnroth, & B. Rydén (Eds.), *Economics and values* (pp. 122-135). Stockholm: Almqvist & Wiksell.

Cultural differences in teaching and learning. *International Journal of Intercultural Relations, 10*(3), 301-320 (SIETAR Award, 1987).

1987

Gevolgen van het Nederlanderschap: Gezondheid, Recht en Economie [*Dutch culture's consequences: Health, law and economy*]. Inaugural Address, University of Limburg, Maastricht. (English translation: Research Memorandum RM 87-037, Department of Economics and Business Administration, University of Limburg at Maastricht.)

The cultural context of accounting. In B.E. Cushing (Ed.), *Accounting and culture* (pp. 1-11). Proceedings of the 1986 Annual Meeting, American Accounting Association, New York.

With J.F.A. Spangenberg. Measuring individualism and collectivism at occupational and organizational levels. In C. Kagitcibasi (Ed.), *Growth and progress in cross-cultural psychology* (pp. 113-122). Lisse, NL: Swets & Zeitlinger.

With D.D. Ohayv. Diagnosticering af organisations-kulturer [Diagnosing organizational cultures]. *Harvard Børsen, 25*(Winter), 60-69. (English translation: Working Paper WP 87-1, Institute for Research on Intercultural Cooperation, University of Limburg at Maastricht.)

The applicability of McGregor's theories in South East Asia. *The Journal of Management Development, 6*(3) 9-18.

1988

With M.H. Bond. The Confucius connection: From cultural roots to economic growth. *Organizational Dynamics, 16*(4), 4-21.

1989

Women in management—A matter of culture. *International Management Development Review*, 5, 250-254.

Cultural predictors of national negotiation styles. In F. Mautner-Markhof (Ed.), *Processes of international negotiations* (pp. 193-201). Boulder CO: Westview.

With J.F.A. Spangenberg. Technik der internationalen Vergleiche [Methodology of international comparative research]. In K. Macharzina & M.K. Welge (Eds.), *Händwörterbuch Export und internationale Unternehmung* (pp. 948-963). Stuttgart: Poeschel. (English translation: Research Memorandum RM 86-013, Department of Economics and Business Administration, University of Limburg at Maastricht.)

With M.H. Bond. The cash value of Confucian values. In S.R. Clegg, S.G. Redding, & M. Cartner (Eds.), *Capitalism in contrasting cultures* (pp. 383-390). Berlin: de Gruyter.

Organising for cultural diversity. *European Management Journal, 7*(4), 390-397.

1990

With B. Neuijen, D.D. Ohayv, & G. Sanders. Measuring organizational cultures: A qualitative and quantitative study across twenty cases. *Administrative Science Quarterly, 35*(2), 286-316.

Nationale waarden in verband met medezeggenschap: is een Europese medezeggenschapsregeling mogelijk? [National values regarding codetermination: Will a European directive for codetermination be possible?]. *M & O, Tijdschrift voor Organisatiekunde en Sociaal Beleid, 44*(5), 459-73.

1991

Cultures and organizations: Software of the mind. London: McGraw-Hill. (Paperback edition London: Harper Collins, 1994. Translations: Dutch, Danish, Swedish, Finnish, German, Norwegian, French, Japanese, Korean, Chinese.)

Management in a multicultural society. *Malaysian Management Review, 25*(1), 3-12.

Empirical models of cultural differences. In N. Bleichrodt & P.J.D. Drenth (Eds.), *Contemporary issues in cross-cultural psychology* (pp. 4-20). Amsterdam: Swets & Zeitlinger.

With R.H. Franke & M.H. Bond. Cultural roots of economic performance: A research note. *Strategic Management Journal, 12*, 165-73.

Marketing en cultuur [Marketing and culture]. In *Contacts beyond contracts* (pp. 23-29). Proceedings of Congress of *Marketing Associatie.* Tilburg: University of Tilburg. (English translation: Working Paper WP 90-06, Department of Economics and Business Administration, University of Limburg at Maastricht.)

1992

Cultural dimensions in people management: The socialization perspective. In V. Pucik, N.M. Tichy, & C.K. Barnett (Eds.), *Globalizing management: Creating and leading the competitive organization* (pp. 139-158). New York: Wiley.

La réintégration de l'Europe de l'Est dans la famille des nations [The Re-integration of Eastern Europe in the family of nations]. *Intercultures, 18*, 27-45. (English translation: Working Paper WP 92-3, Institute for Research on Intercultural Cooperation, University of Limburg at Maastricht.)

Innovation as a cultural phenomenon. In *Technology–A servant of man* (pp. 39-46). Copenhagen: Samfundslitteratur.

1993

Cultural constraints in management theories. *The Executive, 7*(1), 81-94. (Foundation for Administrative Research Distinguished International Scholar Address.)

Intercultural conflict and synergy in Europe. In D.J. Hickson (Ed.), *Management in Western Europe: Society, culture and organization in twelve nations* (pp. 1-8). Berlin: de Gruyter.

Images of Europe. Valedictory address University of Limburg, Maastricht. (Reprinted in *Netherlands Journal of Social Sciences, 30*(1), 63-82.)

With M.H. Bond & C.-L. Luk. Individual perceptions of organizational cultures: A methodological treatise on levels of analysis. *Organization Studies, 14*(4), 483-503.

1994

* *Uncommon sense about organizations: Cases, studies and field observations.* Thousand Oaks, CA: Sage Publications.

The business of international business is culture. *International Business Review, 3*(4), 1-14.

Management scientists are human. *Management Science, 40*(1), 4-13.

Business cultures. *The Unesco Courier, 47*(4), 12-16 (translations in various Unesco languages).

With M. Vunderink. A case study in masculinity/femininity differences: American students in the Netherlands vs. local students. In A.M. Bouvy, F.J.R. van de Vijver, P. Boski, & P. Schmitz (Eds.), *Journeys into cross-cultural psychology* (pp. 329-347). Lisse, NL: Swets and Zeitlinger.

The merchant and the preacher—As pictured by Multatuli's Max Havelaar (1860). In B. Czarniawska-Joerges & P. Guillet de Monthoux (Eds.), *Good novels, better management: Reading organizational realities* (pp. 138-153). Chur, Switzerland: Harwood Academic Publishers.

1995

Verzekering als produkt van nationale waarden [Insurance as a product of national values]. In *Onverzeker-baarheid: wie neemt het risico?*, special issue, *Economisch Statistische Berichten*, 4-8.

With J. Soeters & M. Van Twuyver. Culture's consequences and the police: Cross-border cooperation between police forces in Germany, Belgium and the Netherlands. *Policing and Society, 5*, 1-14.

Verzekering als produkt van nationale waarden. In *Onverzekerbaarheid: wie neemt het risico?*, special issue, *Economisch Statistische Berichten*, February, 4-8. [English version: Insurance as a product of national values. *The Geneva Papers on Risk and Insurance, 20*, 423-429.].

With J. Soeters & M. Van Twuyver. Culture's consequences and the police: Cross-border cooperation between police forces in Germany, Belgium and the Netherlands. *Policing and Society, 5*, 1-14.

Multilevel research of human systems: Flowers, bouquets and gardens. *Human Systems Management, 5*, 207-217.

1996

The nation-state as a source of common mental programming: Similarities and differences across Eastern and Western Europe. In S. Gustavsson & L. Lewin (Eds.), *The Future of the Nation-State: Essays on Cultural Pluralism and Political Integration* (pp. 19-48). Stockholm: Nerenius & Santérus and London: Routledge.

NOTES

1. A.M. Kuylaars, *Werk en leven van de industriële loonarbeider, als object van een sociale ondernemingspolitiek*. Leiden: Stenfert Kroese, 1951.
2. See G. Friedmann, *Problèmes humains du machinisme industriel.* Paris: Gallimard, 1946.
3. See F. Herzberg, B. Mausner, and B.B. Snyderman, *The Motivation to Work*. New York: Chapman & Hall, 1959.
4. See A.H. Maslow, *Motivation and Personality*. New York: Harper & Row, 1954.
5. A.D. de Groot, *Methodologie*. The Hague: Mouton, 1961.
6. C. Argyris, *The Impact of Budgets on People*. Ithaca, NY: School of Business and Public Administration, Cornell University, 1952.
7. A.C. Stedry, *Budget Control and Cost Behavior*. Englewood Cliffs, NJ: Prentice-Hall, 1960.
8. See "The Chinese Culture Connection" (a team of 24 researchers orchestrated by M.H. Bond), Chinese Values and the Search for Culture-Free Dimensions of Culture. *Journal of Cross-Cultural Psychology, 18*(2), 143-164, 1987.
9. See T.S. Kuhn, *The Structure of Scientific Revolutions* (2nd ed.). Chicago: University of Chicago Press, 1970.
10. Uichol Kim, H.C. Triandis, C. Kagitçibaşi, S.C. Choi and G. Yoon (Eds.), *Individualism and Collectivism: Theory, Method, and Applications*. Thousand Oaks CA: Sage, 1994; and H.C. Triandis, *Individualism and Collectivism*. Boulder CO: Westview, 1995.

Roots, Wings, and Applying Management and Leadership Principles: A Personal Odyssey

JOHN M. IVANCEVICH

My journey began at South Chicago Community Hospital on the far south side of what is still today my favorite city: Chicago. I was the firstborn son of Anne Oganovich Ivancevich and Milan ("*Mike*" or "*Legs*") Ivancevich. What I have accomplished throughout my life is linked to my family, their support, encouragement, integrity, work ethic, sensitivity toward others, curiosity, and a thirst to seek knowledge. My personal odyssey is so tightly anchored to my family that the achievements, successes, frustrations, setbacks, and tragedies have largely been shared with relatives.

The south side of Chicago was a diverse mixture of ethnic, religious, and philosophic groups and characters. The neighborhood was "multicultural" long before the concept found its way into the 1990s popular press. Irish, Polish, Italian, Mexican-American, African-American, Serb, Croatian, Hungarian, and Lithuanian foods, languages, newspapers, radio programs, music, and friends were a part of my daily fare. Each day was a full 24-hour series of sounds, excitement, friendships, learning, shared dreams, and hard work in the factories, mills, and shops around the neighborhood.

My grandparents, uncles and aunts, and my mother and father lived in apartments (flats) on Commercial Avenue. A steady stream of visitors visited

Management Laureates, Volume 4, pages 123-151.

ISBN: 1-55938-730-0

the apartments. Each visitor shared a host of opinions, remedies for ailments, and philosophical advice. The patter I still recall is how important it was to be educated. Education was emphasized over and over by everyone—especially my parents. I remember vividly when I graduated from John L. Marsh Elementary School and was talking about going to high school. My father said it is better to think beyond Bowen High School and to earn college and graduate degrees. His dream was that each of his four children would be college educated.

GRADE SCHOOL AND HIGH SCHOOL: THE PREPARATION PHASE

The Chicago Public School System does not enjoy a glowing reputation, but I thoroughly enjoyed school. My first nine years of school (kindergarten included) were at John L. Marsh. I enjoyed every day of school because of the learning environment created by outstanding teachers. My classes stressed learning, skill development, and self-motivation. Although we were a rowdy group, discipline was fairly strict. In my case, the disciplinarians were 13 older uncles and my father and mother. This gang of fifteen kept most of my rowdiness in check.

I am one product of the 1940s and 1950s Chicago Public School System who learned and progressed from grade to grade. I graduated elementary school, moved on to Bowen High School and took what was then referred to as a "college preparatory course." At Bowen, I found outstanding teachers and mentors who prepared me very well. I was a good student who apparently showed some potential because a number of teachers began to ask me about attending college. In the late 1950s, not many graduates of the Chicago Public School System went on to college.

In my second year of high school, I decided that attending college was going to be financially challenging, so I sought out and worked a number of jobs in the steel mills, on construction crews, and in retail establishments. Going to school, playing football, basketball, and baseball, and trying to work took every minute of each week. Somehow I still managed to get into trouble occasionally. But for the most part, I read books, visited museums, and enjoyed talking about a host of subjects with adults.

When I was about 15 years old, Joe Valdez, a neighborhood friend and Purdue University graduate, started talking to me about attending his alma mater. Joe was from Eagle Pass, Texas, and found his way to Purdue and then to Chicago. He provided me with literature and a laundry list of reasons why Purdue was the right school for me. My college-bound high school classmates (there were a few) were going primarily to the University of Illinois, Notre Dame, or schools in the Chicago area.

My first visit to a college campus was to the University of Chicago to take college entrance exams. I received a number of football scholarship offers to schools such as Montana State, the University of Idaho, Valparaiso, and Ripon, but I selected Purdue with the intention of earning either a football or basketball scholarship. My football plans were scuttled because I had no guarantee of an athletic scholarship and because I had to work at least twelve hours a day during the summer to be able to pay for school.

PURDUE UNIVERSITY (1957-1961)

I began my college career assuming that my "natural" academic ability and record of scholastic achievements were enough to make progress with ease. After all, I was a quick learner, voracious reader, and hard worker. To my surprise, I had no idea how to prepare for college classes! My study habits were poor and disorganized. I was homesick and missed organized sports. I was lost and floundering in West Lafayette.

Purdue used a "yellow slip" system to notify parents of students who were not progressing satisfactorily. When I arrived home for the Thanksgiving break during my first semester, my father inquired about the array of "yellow slips" (there were four; I took five classes) that had recently arrived. I provided some form of pathetic excuse-laden gibberish. By some miracle, I was able to shift gears and pass every class with mostly B's the first semester. I had received a specific wake-up call to work harder and smarter.

I stumbled and wandered around Purdue taking courses in engineering and industrial management. I wasn't sure whether I wanted or could be a great engineer. Sputnik had recently lifted off and there was a big push in the United States to educate more engineers. The mathematical and analytical part of engineering coursework was challenging, but doable. The spatial and drawing part of Purdue's engineering preparation was a nightmare. Hours were spent attempting to make sense out of gears, spatial drawings, pulley devices, and many other upside down and inside out configurations.

I worked as an industrial engineering trainee during the summer months and, upon graduation, took a job as an industrial engineering analyst at Republic Steel in Chicago. I was still unsure of an Engineering career and, thinking that working in engineering would lead to some career decision, wanted to experiment by taking on a host of interdisciplinary projects at Republic Steel.

Before graduating from Purdue, I considered attending law school or earning an MBA. It was my opinion that an undergraduate degree was going to be so commonplace in a few years that a graduate education would be required for professional job entry. Uncle Sam and the Berlin Crisis altered my immediate post-Purdue plans. Since I took four years of ROTC at Purdue to earn funds to attend school, I had to fulfill my military obligation. I graduated in May 1961, worked at Republic Steel until September 1961, and began what was supposed to be a six-month tour from November 1961 to April 1962. The Berlin crisis of August 1961 changed everything, and my six-month tour turned into two years beginning in September 1961.

THE U.S. ARMY

My introduction to practicing management and attempting to lead was in the U.S. Army. I was a First Lieutenant trained at Aberdeen Proving Grounds and assigned to the 5th Infantry Division at Fort Carson, Colorado. I was placed in charge of 45 men, millions of dollars of equipment and supply parts, and asked to create a combat-ready platoon.

Of course, there is no college class or textbook that prepared me for the army. My personal background, ability to work with diverse people, and acceptance of challenges helped me navigate through the trials and tribulations of being a supply officer in an infantry division. I was fortunate to learn from Sergeant Herschel Gill about respect, leadership, integrity, and creativity. Although Sergeant Gill reported to me on an organization chart, he served as an invaluable mentor. He prodded, guided, and supported me in planning, organizing, and controlling the goals and actions of a platoon. We worked long hours establishing an efficient supply parts distribution system which supported the entire 5th Infantry Division.

The division moved all around the United States perfecting its role as a military support unit. In October 1962, the Cuban Missile Crisis occurred and the 5th Infantry Division began moving to Florida. Preparing for the invasion of Cuba with a young and inexperienced unit is a memorable event in my life. When 45 lives are tightly bound together and leading by example is a necessity, growing up and applying every skill available is required. Although I am not an "armchair warrior," my two years in the U.S. Army mark the beginning of a career in which I not only have taught, written, and spoken about management, but I also have practiced managing others.

My army experience crystallized my plan to attend graduate school. In February-April 1963, I began applying for scholarships and fellowships in a number of MBA programs. I received small scholarship offers from Michigan State and Berkeley. My wife (Diane Murphy) and I decided that when my two army years were completed, I would attend the University of California at Berkeley and earn an MBA. Everything was set until a Purdue professor, Mort Fowler, contacted me via telephone in July 1963 while I was on military field maneuvers in Yakima Valley, Washington, about a scholarship at the University of Maryland. Upon returning home and talking to Mort, the Maryland scholarship sounded more attractive than the financial support offered by Berkeley. Because I had spent time in Maryland and the financial support was better, we made our way to College Park.

THE UNIVERSITY OF MARYLAND (1963-1968)

I arrived at College Park fresh out of field maneuvers and after a number of interviews with a few generals who attempted to persuade me to stay in the army to take a one-year tour in Viet Nam. The University of Maryland is located on a beautiful campus about seven miles from the U.S. capitol and was beginning to hire a cadre of new professors to initiate a strong MBA program and to create a doctoral program. Charles Taff was a key person in my development at Maryland. He provided the scholarship funding and questioned me about my career plans. He also attracted Frank Paine from Stanford, Steve Carroll and Al Nash from Minnesota, John Ryans and Jim Baker from Indiana, Rudy Lamone from North Carolina, Henry Tosi from Ohio State, and Tony Raia from UCLA. Each of these new professors helped shape my career, study habits, and interest in pursuing the doctorate. These professors were responsible for illustrating that an academic career could be productive, enjoyable, and worthwhile.

Maryland's doctoral program was placed in the hands of a relatively new group of energetic professors. I worked closely on teaching, training programs, consulting in a few firms, and writing projects with most of the new professors. These experiences helped shape my ideas and style of working on field research surveys and field experiments. I learned by observing, application, and receiving harsh but fair feedback on my work products, examinations, and presentations.

In addition to professional influence on my personal growth and development in management, organizational behavior, and analytical procedures, a number of classmates played a significant role. Jim Donnelly, Bob Strawser, Jeff Cahill, Lou Rosen, Ted Matthais, Bill Walker, George Marthiniss, Dick Hise, Jan Mucyzk, and Burt Leete were classmates, friends, and teachers. My peer group at Maryland was very cohesive, supportive, and competitive. We literally lived at school almost seven days a week throughout the year, learning, reading, debating, arguing wildly, and intellectually stimulating each other.

In 1965, I had earned enough credits to receive the MBA and was ready to leave Maryland. My wife was pregnant with our second child, Jill (Dan was born in 1964), and it was time to finally begin a career in the "real world." Enter Charles Taff and his persuasive speeches. He asked me to consider staying on for the doctorate, a new D.B.A. degree, which would be administered by the business administration group and not the economics group that administered the Ph.D. Political power struggles had created a plan at Maryland to develop and implement a new D.B.A. program.

After much soul searching and consideration, my wife and I decided to remain and pursue the doctorate. Until this point, I had not even considered becoming a professor. Charles Taff convinced me that I had an ability to teach and conduct research which could be developed further in the new doctoral program. I am amused whenever someone asks me how I ended up a professor. They appear disappointed that I didn't pursue or start with a lifelong career plan.

In the next three years, Jim Donnelly, Jeff Cahill, Bob Strawser, and I became Maryland's first doctoral students. Jeff had a disagreement with a professor over his master's thesis and dropped out. Bob left Maryland ABD and finished while teaching at Penn State. Jim and I—the two guinea pigs—toughed it out and graduated with everything completed in 1968.

The doctoral education we received was personalized and very intense. Although we attended class after class in every business discipline and economics, learning also occurred outside of class. Jim and I prepared mammoth study guides for the written comprehensives. Who could possibly guess how the new professors would test doctoral candidates? Although we worked on research and writing projects with Tosi, Raia, Paine, Ryans, Carroll, Lamone, and Baker, we were absolutely lost. Jim and I then and now believe that the young professors had no idea what they were doing.

Somehow we successfully completed the comprehensives, survived the orals, and completed dissertations. At the same time, looking for a starting position became another new challenge. Maryland's doctoral students had no idea how the market would react to the product. I received a lot of "no thank you" responses from most schools. On the other hand, a number of schools showed interest and I was fortunate to receive a number of competitive offers.

One example of my job search that would be told differently by my professional colleague and friend John Slocum involved Penn State. I was supposed to be interviewed by Rocco Carzo in Washington, DC, for a possible position at Penn State. However, at the last minute, Rocco had to turn the screening interview over to a new assistant professor, John Slocum. John conducted the interview, and I never heard a word from Penn State. Apparently, John was so unimpressed with my potential that he probably informed Rocco to forget about that "Ivancevich" guy from Maryland. John doesn't remember, but I sure do.

John Ryans, one of my Maryland professors, received a call in February 1968 from John Douglas, at the University of Kentucky, seeking two or three new assistant professors. I had narrowed my choice of offers down to the University of Colorado, Notre Dame, or Southern California when John asked me to visit Lexington. I had no interest in Kentucky, but didn't want to disappoint John Ryans, so I took the trip and eventually accepted the position.

UNIVERSITY OF KENTUCKY (1968-1974)

I arrived in Lexington, Kentucky, with my own plan to teach effectively, conduct rigorous research, create a textbook, and contribute whatever talents I possessed to my students and the state of Kentucky. I planned to be a "professor for all seasons." The new team brought to Lexington in 1968 included Jim Donnelly, Pete Lyon (Illinois), and Andy Grimes (Wisconsin). John Douglas, the chairperson, had assembled what he considered to be an energetic team that could propel Kentucky into a more visible and higher status in academe. John was the Chair, but he was able to recruit this new class because of Dean Chuck Haywood. Chuck was an economist who allowed "green" assistant professors to grow and develop. He provided me with the type of autonomy and academic culture I needed to succeed and fail during my six years at Kentucky. He served as the role model for my own seven years as a dean at the University of Houston.

Besides the incoming team, Kentucky possessed some intellectually challenging and hard working colleagues including Jim Gibson, Stu Klein, Phil Berger, Stu Greenbaum, Jim Knoblett, Joe Massie, and Don Madden. Over the course of my stay at Kentucky, I had the privilege of helping recruit and working with Marc Wallace, Mike Etzel, Bruce Walker, Lynn Spruill, Relmond Van Daniker, and Bill Sartoris. Each of these colleagues provided me with many lessons and insights about teaching, research, and colleagueship.

I became involved in a large goal-setting project while at Maryland and became fascinated with the concept of goals. Because I practiced goal setting daily, I was intrigued by its motivational potential in organizational settings. The field research work and training illustrated to me that practicing managers were attracted to a motivational approach that could be applied across individuals, units, and projects. The Maryland-based work on goal setting served as an important phase in my becoming involved in numerous field surveys and field experiments throughout my career.

Stu Klein, Andy Grimes, John Douglas, and I became involved in a number of interesting field projects at Kentucky Fried Chicken and General Telephone Company. These projects involved goal setting, training, and providing survey feedback to managers. We all seemed to enjoy the research team approach, as well as being involved in attempting to apply what theory and empirical laboratory research suggested in the literature. Kentucky also provided a platform for working on textbook projects that have proven to the interesting, rewarding, and satisfying. Jim Donnelly and I worked on a small workbook project while doctoral students at Maryland. We also taught management courses with a set of notes we had personally developed. At the time we taught, there were only a handful of management texts available, and we believed that our notes would better serve us as young instructors.

The Maryland-developed classroom notes eventually became an idea and the inspiration for teaming up with Jim Gibson at Kentucky and preparing *Fundamentals of Management*, which is now in its 9th edition. Bob Boozer and Jim Sitlington of Business Publications, Inc., found us and our notes and helped us get started in the textbook arena. Bob and Jim, as well as Cliff Francis and Marty Hanifin, are not only publishers, but also close friends. Each of these individuals has illustrated that friendship is more significant than any set of business transactions. Bob, Jim, Cliff, and Marty are classic examples of how academics and practitioners can work effectively as a team. I owe a lot of my textbook accomplishments to these four friends.

I have found that writing, revising, and improving textbooks requires tremendous discipline, constant attention, and careful classroom testing. The Donnelly, Gibson, and Ivancevich team has, fortunately, become one of the most successful in management textbook history. We still remain friends, work on projects, and support each other. Working with colleagues for over 25 years is a testimony to the dedication of each of us to textbook excellence and respect for each other. I learned firsthand the importance of teamwork in writing, revising, testing, and evaluating our work. My academic career would not be as successful and rewarding if Jim Donnelly and Jim Gibson hadn't come along.

One notable event occurred at Kentucky when a senior professor heard that Jim Donnelly and I were working on a textbook as assistant professors. The professor approached me and hinted that textbook work was not how I should proceed while striving to become a tenured associate professor. We debated his stated position, and eventually I informed him that the advice was noted. A few more probes by the senior professor and a few more stringent comments by me finally pointed out that pseudo-threats, advice, or knee-jerk opinions were not going to influence my intentions and goals regarding textbooks. I listened politely but was strong willed about my freedom to choose. The professor and I are still friends, and he probably never recalls the discussions.

My encounter with the senior professor at Kentucky provided me with an important citizenship principle: I never offer advice to a junior faculty member about textbook work, teaching style, service activities, or research projects unless I am asked for an opinion. Each person has his or her own style, work ethic, goals, and threshold for working on multiple projects. I believe that crusty, know-it-all, and opinionated senior professors often possess a large degree of jealousy and envy toward junior professors.

Minding my own business is a rule that I have followed since growing up in the streets of Chicago. In my neighborhood, I observed too many busybodies who were experts about everything, attempting to rearrange their faces after an altercation. Although academics like to consider themselves to be elite and more civilized than neighborhood people, I still believe that each person should pay attention to his or her own situation or be prepared to suffer the consequences.

In 1971, tragedy struck hard when my wife, Diane, passed away ten days after giving birth to our third child, Dana Louise. Textbooks, teaching, research, and career progression were not equivalent to caring for three wonderful children ages seven, five, and ten days old as a single parent. I have never recovered from the shock and suddenness of this tragedy. I now take nothing for granted, and my family is my major joy. Our family circle of five was reduced by the loss of a wonderful wife and mother, but the four of us stayed close to each other.

The loss of my wife created a need to leave Kentucky and move to another career or school. I quietly searched and eventually started to find some possible academic opportunities. Fortunately, I also found a wonderful woman, Margaret (Pegi) Karsner, whom I married in Versailles, Kentucky. Pegi was willing to marry a widower with three young children who would have her leave her beloved Kentucky home. I suffered a tragedy and somehow rebounded to find a miracle.

We decided to leave Kentucky in July 1974. We headed for Houston, and began what was to be the next chapter in a rewarding career and a life sprinkled with disastrous tragedies. I was approached by Mike Matteson and Tim McMahon about joining the University of Houston faculty. Because Houston was in Texas and I had no interest in becoming a Texan, I am still surprised that Mike and Tim convinced me to visit the school.

UNIVERSITY OF HOUSTON (1974-Present)

My career at Houston has been filled with research work, administrative positions, a distinguished professorship, and the major tragedy of my life. I continued work on goal setting and became completely immersed in theoretical model development, empirical studies, and training in the areas of job-related stress and reward system changes in organizations. There have been so many rewarding opportunities to conduct exciting research in a host of organizations while at Houston that I can't imagine a better setting for a management professor. I have been unable to conduct every available research study because of project overload in Houston. In addition, I have given away six or seven major data sets on projects because of other higher priority commitments.

I have enjoyed working with and have learned from a wide array of colleagues while at Houston. My colleagues have included Mike Matteson, Tim McMahon, Jim Phillips, Sara Freedman, Jon Goodman, Jim Sowers, Ken Rediker, Pete Lyon, George Gamble, John Zuckerman, Benton Cocanougher, Bob Keller, Mary-Jane Saxton, Gail Arch, Ron Salazar, Dick DeFrank, Art Jago, Dave Schweiger, Ev Gardner, Joanne Verdin, Kim Stewart, Jackie Gilbert, Tom Duening, Roger Blakeney, Dutch Holland, Dick Montanari, Fred Dorin, Rich Arvey, Jim Terborg, Bob Pritchard, and many others. Each of these and numerous other colleagues have taught me about teaching, research, service, and colleagueship.

Long before it became fashionable in the popular press, I studied and wrote about dysfunctional job stress and how expensive it is personally and organizationally. Mike Matteson and I teamed up for years studying, analyzing, and intellectually debating about workplace stress. Research projects, radio and television shows, courses, books, cassette tapes, training programs, collaborative projects with medical researchers, and popular press commentary all resulted from our immersion in the stress area. We examined sources of stress, individual reactions to stress, and organizational consequences of workplace stress in a host of institutions. Our diagnostic techniques included surveys, interviews, and use of biochemical measures. We always attempted to link our work to possible intervention approaches that could reduce stress.

The notion of a healthy organization still intrigues me, and someday policymakers will realize that preventive health models are more feasible for society than medical health models or approaches. As U.S. health-care expenditures continue to skyrocket, the emphasis on preventive approaches still lags miserably in the American psyche. Individuals, families, and organizations need to make health promotion and prevention of premature illness and disease the very top priority in quality of life improvement.

The stress stream of research and other outcomes was very intrinsically rewarding. Mike and I experimented with many different types of intervention. Unfortunately, being a hard-core Type A precluded me personally from having much of a chance to reduce my own stress levels. In this case, I know a lot about stress mechanisms in the body, but I do not practice sound stress reduction or prevention techniques. The uncontrollable environmental forces and my own makeup constantly resist any practice of good prevention and stress management. The 24-hour day needs to be extended to at least 48 hours per day for me to be able to practice what I know is best.

Houston also provided me with numerous opportunities to conduct organizational change interventions and evaluations. I worked in many industries and in institutions of all sizes. Evaluating change longitudinally and dealing with boards of directors or company research councils kept me busy and also sharpened my research and presentation skills. Convincing any committee that was adamantly opposed to most research that a particular study and design is best required a lot of planning, cajoling, and commitment. On more than a few occasions, the plan of the research project was abruptly changed or aborted in the middle of an exciting field experiment. Attempting to salvage a longitudinal, tightly designed project is difficult, to say the least. I was occasionally able to hold the line, but in so many cases, a study simply evaporated. I was fortunate to have a number of years' experience before my first episode of changing or stopping a study occurred. Even a person with some research scar tissue can become frustrated and upset with the field research world.

To have a full range of field and laboratory research experience, I purposefully initiated or became involved with lab studies. I wanted to feel and experience the difference in controlling the environment. I found that lab research, like field research, has its pluses and minuses. I also concluded that lab researchers need to get out into the field and vice versa. Staying within a single research domain is not what I would recommend to any young scholar. Feel it, try it, and experience it in its totality is my unsolicited advice. Houston was the place that permitted me to move in and out of lab and field settings. The only thing that has stopped me from conducting rigorous and tightly designed research has been serving as a dean. Even I couldn't squeeze in the needed quality time to conduct solid empirical research on all of the problems of interest. What non-researchers and a lot of critics of research fail to understand is that a well-constructed longitudinal study and report requires hundreds of hours and intense discipline to conduct each step properly.

I have served as the chairperson of the Department of Management, the Associate Dean for Research, and the Dean while at Houston. In each of these assignments, I have attempted to practice what I teach, write, and research—managing and leading people. My administrative roles provided me with many uplifting experiences, some frustrations, and a few disappointments. An example of an uplifting experience was, in 1976, having a major role, as a chairperson, in recruiting Sara Freedman (North Carolina), Art Jago (Yale), and Dick Montanari (Colorado). Each of these colleagues has turned out to be not only exceptional academics but also tremendous human beings. Although Dick and Art have moved on and Sara is still at Houston, they represent what I believe are well-rounded, contributing academic role models. Each of these scholars has improved the stature of the field of business education.

I have been frustrated by the lack of interest displayed by too many faculty colleagues in the state of business education. Unless academics begin to realize what customers need, what society requires, and what legislators almost demand, I am afraid that business education will begin to lose its vitality and stature in academe. Serving as a dean has broadened my perspective about education and how it must change.

My disappointments have largely centered on broken promises made by administrators and faculty members to complete a project. I am very impatient and almost intolerant with promise breakers. One of the reasons I have come to enjoy Texans is that a handshake or a person's word serves as a promise. Native Texans are "can do" people, and, despite all of the negative national press hype, if I am in a bind, I'll seek out a Texan to stay the course and help solve the problem.

The most significant position I have held at Houston since 1979 is the Hugh Roy and Lillie Cranz Cullen Professor of Management. The Cullen Distinguished Professorships have been the most academically prestigious positions in the entire institution. There are about 15 Cullen Professors among a faculty totaling around 1,000. In accepting the Dean's position in 1988, I insisted on retaining the Cullen position while serving and after completing my term as dean (1988-1995). I take serving as an active distinguished professor of management seriously. It is not just a title; rather, it is an action-oriented position that serves every area and discipline of the college. Conducting research, developing courses and programs, teaching, providing service professionally, locally, across the campus, and in the college, and helping junior faculty members achieve their goals are what a Cullen Professor provides.

While serving as the dean, I purposefully attempted and succeeded in teaching at least one undergraduate course every year and publishing every year. Too many administrators brag about not setting foot in a classroom for years and that publishing anything is impossible because of the lack of time. In my opinion, every administrator, including the chancellor, should sign an oath that he or she will teach a course every year they serve. Furthermore, every administrator needs to publish something of value as much as possible since creating knowledge is the essence of being a scholar in higher education.

I learned more about what was occurring in the College by teaching juniors and seniors than by any other method. Students were surprised to learn that "the dean" was interested and dedicated to teaching, listening, and learning. By listening and learning, I was able to initiate a number of student service changes that the College should be able to use for many more years.

I forced myself to create new text products, working with Mike Matteson on *Organizational Behavior and Management*; with Steve Skinner, Peter Lorenzi, and Phil Crosby on *Management: Quality and Competitiveness*; and with Steve Skinner on *Business in the 21st Century*. The *Management* book was very interesting, because it allowed me to work with two enjoyable and knowledgeable academics—Steve and Peter—and with Phil Crosby, the quality guru. Planning the book, visiting Phil's Florida home, and working with the publisher to produce a different kind of management book proved to be interesting. Trying to produce new pedagogical packages and working with colleagues from many different backgrounds has meant that I am continually learning.

Since arriving in Houston in 1974, I have been able to travel around the world. I have become an enthusiastic supporter of internationalism and have attempted to instill within the culture of the College and in the University a genuine interest in studying and learning about our interdependent global world. Visits around the globe, international visitors to the University and the College, and working on international research projects have reinforced my view that Americans need to become more globally connected. My most rewarding international work was in Madrid, Spain, from 1988-1994 creating and administering the Madrid Business School. Art Jago, Tim McMahon, and I struggled, debated, sweated, and worked with Spanish colleagues to initiate a high quality MBA program. Spanish politics and financial exigencies, however, brought the school to a halt. Despite the disappointing demise of the school, I learned firsthand about international cooperation, negotiation, and conflict, as well as about national cultural differentials.

After serving for seven years as a dean, I decided to step out of administration. A complete administrative shakeup at Houston in which the Chancellor, President, and Provost all resigned within three weeks resulted in my being asked to become the Provost in a new regime. After several weeks of careful analysis, career evaluation, and personal considerations, I decided to accept the Provost job and am looking forward to serving well in the role. It was not a job or position I ever sought, and psychologically taking the Houston Provost position was never a part of my career goals. Surprisingly, many colleagues indicated that rejecting the Provost position would mean that I would not be in a position that involved action. I have been able over the course of my career to create my own action including research projects, new classes, textbook writing, speaking engagements, consulting projects, and training presentations. Advising me that I needed an administrative role to be in the action zone indicates that the friendly advisors have no knowledge about who I am or where I came from.

I am reluctant to share my personal grief, but want to state that in December 1992 our twenty-one-year-old daughter, Dana Louise, passed away unexpectedly. This was a numbing blow from which I will never recover emotionally. Time has not healed, nor will it ever heal my sense of loss. Dana was our special sun, mountain, and water whom I miss every waking moment. Losing a wife and a daughter, and few other personal problems have forced me to consider many aspects of life, career, friendship, and family. Losing a child in the prime of her life has clearly signaled my own mortality and taken away a part of my enthusiasm, zest for life, and upbeat spirit.

CONCLUDING REMARKS

In reviewing my awkward exposition, I realize that I have revealed too much about my personal life. However, it is difficult to determine the feelings, character, and style of a man or woman without knowing something about where he or she came from and what roads he or she has traveled. My roots were in a multicultural, lower middle-class ghetto in Chicago. I didn't know how poor or ghetto-like my early years were until I read a number of University of Chicago sociological analyses about my neighborhood. The University of Chicago sociologists were so off base with assumptions and conclusions that

I enjoyed many good laughs reading what to me were inaccurate comments about neighborhoods and ethnic groups. I wasn't poor in terms of family values, parental and family love, friendships, laughter, and a sense of community. I reflect on my childhood and find pearls of wisdom, philosophy, and intellectual stimulation everywhere. I found out early that steel-mill workers, single-parent mothers of large families, shop owners, immigrants, clergymen, professional fighters, military veterans, school teachers, ex-convicts, and hundreds of others I lived with in Chicago have ideas, opinions, stories, and insights that are invaluable.

I learned that intellectual input and sharing doesn't only come from working as an academic in a university or college. Intellectual debate and curiosity were always a part of my childhood. Upon reflection, it now seems likely that I would be attracted to a formal academic life that encourages intellectual debate and curiosity. My upbringing among mostly non-college role models didn't dampen my enthusiasm for learning. I applaud and thank my childhood mentors for nurturing strong family roots and then pushing me to use my wings to fly off to college and a world a great distance from our south side apartment. My own children still cannot comprehend that my Chicago family of six lived in a flat that was about 700 square feet. Space wasn't that important, but sharing, working together, supporting each other, and keeping promises were the order of each day.

I was able to study management, research organizations, and attempt to lead men and women because of the type of childhood I enjoyed. There is no economic value that can be applied to upbringing and family. In my case, I was rich beyond all comprehension. Uncles, aunts, cousins, grandparents, parents, a tremendous brother, Steve, two wonderful sisters, Donna and Georgene, and hundreds of friends and neighbors. What else does one need to become a self-actualized adult?

In working with thousands of students over my career, I only hope that each one received something of value from our acquaintance. I am who I am because of where I came from. Chicago, Purdue, the U.S. Army, Maryland, Kentucky, and Houston have all been a part of my journey. Each place and every person I have worked with or shared life with has helped me develop as a person. I dedicate this essay to my family—Dan, Jill, Dana, Pegi, and Diane—and to each person who touched my life.

PUBLICATIONS

1966

With J.H. Donnelly, Jr. Steps toward professionalization of training directors. *Personnel Journal*, December, 662-668.

With J.H. Donnelly, Jr. *Elements of business enterprise workbook. Ronald Press.*

1967

With James H. Donnelly, Jr. The current state of bank marketing. *Business and Economic Dimensions*, January, 3-4, 27-32.
Improving business school personnel courses. **Improving College and University Teaching**, Winter, 54-56.
With J.H. Donnelly, Jr. Small banking marketing strategies. **Journal of Small Business Management**, July, 10-13.

1968

With J.H. Donnelly, Jr.& R.T. Hise. Have commercial banks adopted the marketing concept? *Business Studies*, Fall, 172-175.
With J.H. Donnelly, Jr. Current developments in bank marketing. *Mississippi Valley Journal of Business and Economics*, Fall, 46-53.

1969

With R.H. Strawser. A comparative analysis of the job satisfaction of industrial managers and certified public accountants. *Academy of Management Journal*, May, 193-203.
Perceived need satisfactions of domestic versus overseas managers. *Journal of Applied Psychology*, June, 274-278.
With R.H. Strawser & H.L. Lyon. A note on the job satisfaction of accountants in large and small CPA firms. *Journal of Accounting Research*, Autumn, 339-345.
Selection of American managers for overseas assignments. *Personnel Journal*, March, 189-193, 200.
With J.C. Baker. The job satisfaction of American managers overseas. *MSU Business Topics*, August, 72-78.
With J.C. Baker. A study of the reasons for on-the-job performance failures of American expatriates. *Business Review*, Winter, 42-49.
With J.H. Donnelly, Jr. How marketing oriented are commercial banks? *Journal of the American Bankers Association*, February, 59-60.
Predeparture training for overseas: A study of American manager training for overseas transfer. *Training and Development Journal*, February, 36-40.
The theory and practice of management by objectives. *Michigan Business Review*, March, 13-16.

1970

With J.H. Donnelly, Jr. *Analysis for marketing decisions.* Homewood, IL: Richard D. Irwin, Inc.

With J.K. Ryans & J.H. Donnelly, Jr. *New dimensions in retailing: A decision-oriented approach.* Belmont, CA: Wadsworth Publishing Co.

With J.C. Baker. A comparative study of the satisfaction of domestic United States managers and overseas United States managers. *Academy of Management Journal*, March, 69-77.

With J.H. Donnelly, Jr. Post-purchase reinforcement and back-out behavior. *Journal of Marketing Research*, August, 399-400.

With J.H. Donnelly, Jr. Study of consumer political orientations and store patronage. *Journal of Applied Psychology*, October, 470-472.

With J.H. Donnelly, Jr. Leader influence and performance. *Personnel Psychology*, Winter, 539-549.

An analysis of control, bases of control, and satisfaction in an organizational setting. *Academy of Management Journal*, December, 427-436.

With J.C. Baker. Multinational management staffing with American expatriates. *Economic and Business Bulletin. Temple University*, Fall, 35-39.

With H.L. Lyon. Dummy variables: A practical aid in sales forecasting. *Journal of Business Administration*, Winter, 43-51.

With J.H. Donnelly, Jr. & H.L. Lyon. A study of the impact of management by objectives on perceived need satisfaction. *Personnel Psychology*, pp. 139-151.

1971

Fundamentals of management: Functions, behavior, models. Plano, TX: Business Publications, Inc.

Fundamentals of management: Selected readings. Plano, TX: Business Publications, Inc.

With H.L. Lyon & J.H. Donnelly, Jr. A motivational profile of the management scientist. *Operations Research*, October, 1282-1299.

With R.H. Strawser. Job satisfactions of accountants. *GAO Review*, Spring, 28-36.

With J.C. Baker. The assignment of American executives abroad: Systematic, haphazard, or chaotic? *California Management Review*, Spring, 39-44.

With J.H. Donnelly, Jr. Job offer acceptance behavior and reinforcement. *Journal of Applied Psychology*, April, 119-122.

1972

A longitudinal assessment of management by objectives. *Administrative Science Quarterly*, March, 126-138.

With S.J. Carroll, Jr. & F.T. Paine. The relative effectiveness of training methods—Expert opinion and research. *Personnel Psychology*, Fall, 495-509.

1973

Organizations: Behavior, structure, processes. Plano, TX: Business Publications, Inc.

Readings in organizations. Plano, TX: Business Publications, Inc.

With P.K. Berger. Birth order and managerial achievement. *Academy of Management Journal*, September, 515-519.

A progression training model for management by objectives. *Training and Development Journal*, September, 24-30.

With J.H. Donnelly, Jr. & L.A. Capaldini. Operations research and model building. In S.H. Britt (Ed.), *Marketing handbook* (Ch. 24). Dartnell Corporation.

1974

A study of a cognitive training program: Trainer styles and group development. *Academy of Management Journal*, September, 428-439.

With J.H. Donnelly, Jr. A methodology for identifying innovator characteristics of new brand purchases. *Journal of Marketing Research*, August, 331-334.

With M.J. Etzel. Management by objectives in marketing: Philosophy, process and problems. *Journal of Marketing*, October, 47-55.

Changes in performance in management by objectives program. *Administrative Science Quarterly*, December, 563-574.

With H.L. Lyon. An exploratory investigation of organization climate and job satisfaction in a hospital. *Academy of Management Journal*, December, 635-648.

Effects of the shorter workweek on selected satisfaction and performance measures. *Journal of Applied Psychology*, December, 717-721.

With J.H. Donnelly, Jr. A study of role clarity and need for clarity in three occupational groups. *Academy of Management Journal*, March, 28-36.

With M.J. Etzel, H.L. Lyon, & B.J. Walker. A modified nominal-group process for public-sector problem solving. *Public Personnel Management*, September-October, 439-446.

1975

Fundamentals of management: Functions, behavior, models (2nd ed.). Plano, TX: Business Publications, Inc.

Fundamentals of management: Selected readings (2nd ed.). Plano, TX: Business Publications, Inc.

With J.H. Donnelly, Jr. Relation of organizational structure to job satisfaction, anxiety-stress, and performance. *Administrative Science Quarterly*, June, 272-280.

With J.H. Donnelly, Jr. Role clarity and the salesman. *Journal of Marketing*, January, 71-74.

With M.J. Wallace, Jr. & H.L. Lyon. Measurement modifications for assessing organizational climate in hospitals. *Academy of Management Journal*, March, 82-97.

With J.H. Donnelly, Jr. & J.L. Gibson. Evaluating MBO: The challenge ahead. *Management by Objectives*, pp. 15-23.

1976

Organizations: Behavior, structure, processes (2nd ed.). Plano, TX: Business Publications, Inc.

Readings in organizations (2nd ed.). Plano, TX: Business Publications, Inc.

General business: Concepts, values, skills (H.L. Lyon, J.M. Ivancevich, & D.H. Kruger). New York: Harcourt, Brace, Jovanovich.

Management science in organizations (H.L. Lyon, J.M. Ivancevich, & J.H. Donnelly, Jr.). Goodyear Publishing Co.

With M.J. Etzel & J.H. Donnelly, Jr. Social character and consumer innovativeness. *Journal of Social Psychology*, December, 153-154.

Effects of goal setting on performance and job satisfaction. *Journal of Applied Psychology*, October, 605-612.

With J.T. McMahon. Group development, trainer style, and carry over job satisfaction and performance. *Academy of Management Journal*, September, 395-412.

With J.T. McMahon. A study of control in a manufacturing organization: Managers and nonmanagers. *Administrative Science Quarterly*, March, 66-83.

Predicting job performance by use of ability tests and studying job satisfaction as a moderating variable. *Journal of Vocational Behavior*, August, 87-97.

With M.T. Matteson & J.T. McMahon. Organizational climate and job satisfaction of medical technologists. *American Journal of Medical Technology*, October, 15-20.

Expectancy theory predictions and behaviorally anchored scales of motivation: An empirical test of engineers. *Journal of Vocational Behavior*, March, 59-75.

1977

With A.D. Szilagyi, Jr. & M.J. Wallace, Jr. *Organizational behavior and performance.* Goodyear Publishing Co.
With A.D. Szilagyi, Jr. & M.J. Wallace, Jr. *Readings in organizational behavior and performance.* Goodyear Publishing Co.
Management classics. Goodyear Publishing Co.
With J.T. McMahon. Black-white differences in a goal setting program. *Organizational Behavior and Human Performance*, December, 287-300.
With J.T. McMahon. A study of task goal attributes, higher order need strength, and performance. *Academy of Management Journal*, December, 552-563.
A multitrait-multirater analysis of a behaviorally anchored rating scale for sales personnel. *Applied Psychological Measurement*, Fall, 523-532.
With J.T. McMahon. Education as a moderator of goal setting effectiveness. *Journal of Vocational Behavior*, pp. 83-94.
With J.T. McMahon & M.T. Matteson. Relationship of individual need satisfaction and organizational practices to job satisfaction among laboratory personnel. *American Journal of Medical Technology*, pp. 15-19.
With H.L. Lyon. The shortened workweek: A field experiment. *Journal of Applied Psychology*, February, 34-37.
Different goal setting treatments and their effects on performance and job satisfaction. *Academy of Management Journal, September, 406-419.*

1978

Fundamentals of management: Functions, behavior, models (3rd ed.). Plano, TX: Business Publications, Inc.
Fundamentals of management: Selected readings (3rd ed.). Plano, TX: Business Publications, Inc.
The performance to satisfaction relationship: A causal analysis of stimulating and non-stimulating jobs. *Organizational Behavior and Human Performance*, December, 350-365.
With M.T. Matteson. Organizations and coronary heart disease: The stress connection. *Management Review*, October, 14-19.
With M.T. Matteson. Longitudinal organizational research in field settings. *Journal of Business Research*, August.
With J.T. McMahon, J.W. Streidl, & A.D. Szilagyi, Jr. Goal setting: The Tenneco approach to personnel development and management effectiveness. *Organizational Dynamics*, Winter, 63-80.
With H.L. Lyon. A behavioral study of nurses, supervisors, and diagnosticians in a hospital setting. *Decision Sciences*, April, 259-272.
With A.B. Cocanougher. A behavioral-based performance rating technique for sales force personnel. *Journal of Marketing*, July, 87-95.

1979

Organizations: Structure, process, behavior (3rd ed.). Plano, TX: Business Publications, Inc.

Developed three cassettes on: (1) Managing stress: An introduction and overview; (2) Work related stress: Sources and solutions; and (3) External stressors and individual programs for managing stress, for the American Management Association, New York.

With H.L. Lyon & D.P. Adams. *Business in a dynamic environment*. St. Paul, MN: West Publishing Co.

Longitudinal study of the effects of rater training on psychometric error in ratings. *Journal of Applied Psychology*, pp. 502-508.

An analysis of participation in decision making among project engineers. *Academy of Management Journal*, June, 253-269.

With M.T. Matteson. Organizational stressors, physiological and behavioral outcome, and coronary heart disease: A research model. *Academy of Management Review*, July, 347-358.

High and low task stimulation jobs: A causal analysis of performance satisfaction relationship. *Academy of Management Journal*, June, 347-358.

With M.J. Etzel. Job enrichment in marketing. *California Management Review*, Fall, 88-95.

With M.T. Matteson. What managers want to know about job stress. *Management World*, July, 4-6.

1980

Managing for performance. Plano, TX: Business Publications, Inc.

With M.T. Matteson. *Stress and work: A managerial perspective*. Glenview, IL: Scott, Foresman & Co.

With M.T. Matteson. Optimizing human resources: A case for preventive health and stress management. *Organizational Dynamics*, Autumn, 5-25.

With R. Arvey. Punishment in organizations: A review, propositions, and research suggestions. *Academy of Management Review*, January, 123-132.

A longitudinal study of behavioral expectation scales: Attitudes and performance. *Journal of Applied Psychology*, pp. 139-149.

With M.T. Matteson. The coronary-prone behavior pattern: A review and appraisal. *Social Science and Medicine*, June, 337-351.

Behavioral expectation scales versus nonanchored and trait rating systems: A salespersonnel application. *Applied Psychological Measurement*, Winter, 131-133.

With M.T. Matteson. Nurses and stress: Time to examine the potential problem. *Supervisor Nurse*, June, 17-22.

With M.T. Matteson & J.T. McMahon. Understanding professional job attitudes. *Hospital and Health Services Administration*, Winter, 53-68.

1981

Fundamentals of management: Functions, behavior, models (4th ed.). Plano, TX: Business Publications, Inc.

Fundamentals of management: Selected readings (4th ed.). Plano, TX: Business Publications, Inc.

Management classics (2nd ed.). Goodyear Publishing Co.

With S.V. Smith. Goal setting interview skills training: Simulated and on-the-job analyses. *Journal of Applied Psychology, 66*, 697-705.

With S.V. Smith. Identification and analysis of job difficulty dimensions: An empirical study. *Ergonomics, 24*, 351-363.

Management and compensation. In *Handbook of organizational quality of life* (pp. 69-85). John Wiley & Sons.

With C.A. Preston & M.T. Matteson. Stress and the OR nurse. *AORN Journal*, March, 662-671.

The psychology of management. *Texas Hospitals*, December, 8-11.

1982

With M.T. Matteson. *Managing job stress and health*. New York: Free Press.

Organizations: Structure, process, behavior (4th ed.). Plano, TX: Business Publications, Inc.

With J.L. Gibson & J.H. Donnelly, Jr. *Readings in organizations* (3rd ed.). Plano, TX: Business Publications, Inc.

Subordinates' reactions to performance appraisal interviews: A test of feedback and goal setting techniques. *Journal of Applied Psychology, 67*, 581-587.

With J.T. McMahon. The effects of goal setting, external feedback, and self-generated feedback on outcome variables: A field experiment. *Academy of Management Journal*, June, 359-372.

With M.T. Matteson. Type A and B behavior patterns and self-reported health symptoms and stress: Examining individual and organizational fit. *Journal of Occupational Medicine*, August, 585-589.

With M.T. Matteson. The how, what and why of stress management training. *Personnel Journal*, October, 768-774.

With M.T. Matteson. Stress and the medical technologist: I. A general overview. *American Journal of Medical Technology*, March, 163-168.

With M.T. Matteson. Stress and the medical technologist: II. Sources and coping mechanisms. *American Journal of Medical Technology*, March, 169-176.

With M.T. Matteson & C.A. Preston. Occupational stress, type A behavior, and physical well being. *Academy of Management Journal*, June, 373-391.
With S.V. Smith. Job difficulty as interpreted by incumbents: A study of nurses and engineers. *Human Relations*, October, 391-412.
With M.T. Matteson. Occupational stress, satisfaction, physical well being and coping: A study of homemakers. *Psychological Reports*, June, 995-1005.

1983

Contrast effects in performance evaluation and reward practices. *Academy of Management Journal*, September, 465-476.
With M.T. Matteson. Note on tension discharge rate as an employee health status predictor. *Academy of Management Journal*, September, 540-545.
With M.T. Matteson. Employee claims for damages add to the high cost of job stress. *Management Review*, November, 9-13.
With H.A. Napier & J.C. Wetherbe. Occupational stress, attitudes, and health problems in the information systems professional. *Communications of the ACM*, October, 800-806.
With J.H. Donnelly, Jr. & L.A. Capaldini. Management science and model building. In *Marketing manager's handbook*. Chicago: Dartnell Corp.
Managing for performance. Plano, TX: Business Publications, Inc.
With W.F. Glueck (deceased). *Foundations of personnel/human resource management*. Plano, TX: Business Publications, Inc.
With H.L. Lyon & D.P. Adams. *Introduction to business*. St. Paul, MN: West Publishing Co.

1984

Fundamentals of management (5th ed.). Plano, TX: Business Publications, Inc. (Adopted by over 300 institutions in 4th ed.).
Perspectives on management (5th ed.). Plano, TX: Business Publications, Inc.
With M.T. Matteson & S.V. Smith. Relation of type A behavior to performance and satisfaction among sales personnel. *Journal of Vocational Behavior*, 203-214.
With M.T. Matteson. A type A-B person-work environment interaction model for examining occupational stress and consequences. *Human Relations*, pp. 491-513.
Management in the future. A chapter for the *Encyclopedia of professional management*. New York: McGraw-Hill.

1985

Organizations: Behavior, structure, processes (5th ed.). Plano, TX: Business Publications, Inc. (Adopted by over 295 institutions in 4th ed.).

Organizations: Close-up (5th ed.). Plano, TX: Business Publications, Inc.

With R.S. DeFrank, M.T. Matteson, & D.M. Schweiger. The impact of culture on the management practices of American and Japanese CEOs. *Organizational Dynamics*, Spring, 62-76.

With M.T. Matteson and E.P. Richards, III. Who's liable for stress on the job? *Harvard Business Review*, March-April, 60-62, 66, 70-72.

Predicting absenteeism from prior absence and work attitudes. *Academy of Management Journal*, March, 219-228.

With A. Napier & J.C. Wetherbe. An empirical study of occupational stress, attitudes and health among information systems personnel. *Information & Management*, September, 77-85.

With D.L. Schweiger. Human resources: The forgotten factor in mergers and acquisitions. *Personnel Administrator*, November, 47-48, 50-51, 53, 58-61.

1986

Foundations of personnel/human resource management (3rd ed.). Plano, TX: Business Publications, Inc.

With J.H. Donnelly, Jr. & J.L. Gibson. *Managing for performance* (3rd ed.). Plano, TX: Business Publications, Inc. (Adopted by over 200 institutions in 2nd ed.).

With M.T. Matteson. *Management classics* (3rd ed.). Plano, TX: Business Publications, Inc.

With D.M. Schweiger & F.R. Power. Executive actions for managing human resources before and after acquisitions. *Academy of Management Executive*, May, 127-138.

With M.T. Matteson. Organizational level stress management interventions: A review and recommendations. *Journal of Organizational Behavior Management*, Fall/Winter, 229-248.

Life events and hassles as predictors of health symptoms, job performance, and absenteeism. *Journal of Occupational Behavior*, January, 39-51.

With R.S. DeFrank. Job loss: An individual level review and model. *Journal of Vocational Behavior*, February, 1-20.

With M.T. Matteson. Medical technologists and laboratory technicians: Sources of stress and coping strategies. In *Stress in the health professions*. New York: John Wiley & Sons.

With M.T. Matteson. An exploratory investigation of CES as an employee stress management procedure. *Journal of Health and Human Resources Administration*, Summer, 93-109.

With M.T. Matteson. Organizational change may be hazardous to your health. *Corporate Commentary*, Spring/Summer, 18-26.

1987

With M.T. Matteson. *Organizational behavior and management*. Plano, TX: Business Publications, Inc.

With J.H. Donnelly, Jr. & J.L. Gibson. *Fundamentals of management* (6th ed.). Plano, TX: Business Publications, Inc. (Adopted by over 300 institutions in 5th ed.).

With J.H. Donnelly, Jr. & J.L. Gibson. *Perspectives on management* (6th ed.). Plano, TX: Business Publication, Inc.

With D.C. Ganster. *Job stress: From theory to suggestions*. New York: Haworth Press.

With M.T. Matteson. *Controlling work stress*. San Francisco: Jossey-Bass.

With C. Leana. Involuntary job loss: Institutional interventions and an agenda for research. *Academy of Management Review*, April, 301-312.

With S.J. Skinner & J.H. Donnelly, Jr. Effects of transactional form on environmental linkages and power-dependence relations. *Academy of Management Journal*, September, 577-588.

With M.T. Matteson & G.O. Gamble. A test of the cognitive social learning model of type A behavior. *Journal of Human Stress*, Spring, 23-31.

With M.T. Matteson. Individual stress management intervention: Evaluation of techniques. *Journal of Managerial Psychology*, June, 24-30.

With M.T. Matteson & G.O. Gamble. Birth order and the type A coronary behavior pattern. *Individual Psychology*, March, 42-49.

With D.M. Schweiger & F.R. Power. Strategies for managing human resource during mergers and acquisitions. *Human Resource Planning*, June, 19-36.

With M.T. Matteson. Medical technologists and laboratory technicians: Sources of stress and coping strategies. In R. Payne & J. Firth-Cozene (Eds.), *Stress in health professionals* (pp. 231-256). New York: John Wiley.

With M.T. Matteson. Worker health and type A organizations. *Business and Health*, October, 12-16.

1988

With J.L. Gibson & J.H. Donnelly, Jr. *Organizations: Behavior, structure, processes* (6th ed.). Plano, TX: Business Publications, Inc. (Adopted by over 325 institutions in 5th ed.).

With J.L. Gibson & J.H. Donnelly, Jr. *Organizations: Close-up* [*6th ed.*]. *Plano, TX: Business Publications, Inc.*

With R. DeFrank & D.M. Schweiger. Job stress and mental well being: Similarities and differences among American, Japanese, and Indian managers. Behavioral Medicine, Winter, 160-170.

With M.T. Matteson. Type A behavior and the healthy individual. *British Journal of Medical Psychology*, March, 37-56.

With M.T. Matteson. Health promotion at work. *International Review of Industrial and Organizational Psychology*, pp. 279-306.

With M.T. Matteson. Promoting the individual's health and well being. In C.L. Cooper & R. Payne (Eds.), *Stress at work* (pp. 267-299). New York: John Wiley.

With M.T. Matteson. Worksite health promotion: Some important questions. *Health Values*, January/February, 23-29.

With D.M. Schweiger & F.R. Power. Human resource management: A forgotten merger and acquisition consideration. In R.S. Schuler & S.A. Youngblood (Eds.), *Readings in personnel and human resource management* (pp. 92-107). St. Paul: West. (Article is similar to original paper that appeared in *Human Resource Planning*, June 1987 [see full reference below]).

With M.T. Matteson. Application of the triangulation strategy to stress. In J.J. Hurrell, Jr., L.R. Murphy, S.L. Sauter, & C.L. Cooper (Eds.), *Occupational stress: Issues and development in research* (pp. 200-215).

1989

With J.H. Donnelly, Jr. & J.L. Gibson. *Management: Principles and functions* (4th ed.). Homewood, IL: Richard D. Irwin, Inc. & Plano, TX: Business Publications, Inc.

Foundations of personnel/human resource management (4th ed.). Homewood, IL: Richard D. Irwin, Inc. & Plano, TX: Business Publications, Inc.

With M.T. Matteson. *Management classics* (4th ed.). Homewood, IL: Richard D. Irwin, Inc. & Plano, TX: Business Publications, Inc.

With K. Stewart. Appraising management talent in acquired organizations: A four-tiered recommendation. *Human Resource Planning*, May, 141-154.

1990

With J.H. Donnelly, Jr. & J.L. Gibson. *Fundamentals of management* (7th ed.). Homewood, IL: Richard D. Irwin, Inc.

With M.T. Matteson. *Organizational behavior and management* (2nd ed.). Homewood, IL: Richard D. Irwin, Inc.

With M.T. Matteson, S.M. Freedman, & J.S. Phillips. Work site stress management interventions. *American Psychologist*, February, 252-261.

With J.S. Phillips, S.M. Freedman, & M.T. Matteson. Type A behavior, self-appraisals, and goal setting: A famework for future research. *Journal of Social Behavior and Personality*, June, 59-76.
With M.T. Matteson. Merger and acquisition stress: Fear and uncertainty at mid-career. *Prevention In Human Services, 8*(1). (Also appears in: J.C. Quick, R.E. Hess, J. Hermalin, & J.D. Quick (Eds.), *Career stress in changing times*, New York: Haworth Press.)

1991

A traditional faculty member's perspective on entrepreneurship. *Journal of Business Venturing*, January, 1-7.
With J.L. Gibson & J.H. Donnelly, Jr. *Organizations: Behavior, structure, processes* (7th ed.). Homewood, IL: Richard D. Irwin, Inc. (Adopted by over 300 institutions in 7th ed.).

1992

With R.S. DeFrank & P.R. Gregory. The Soviet enterprise director: An important resource before and after the coup. *Academy of Management Executive*, February, 42-55.
With J. Quick et al. Occupational mental health promotion: A prevention agenda based on education and treatment. *American Journal of Health Promotion*, September-October, 37-44.
With S. Skinner. *Business in the 21st century*. Homewood, IL: Richard D. Irwin, Inc.
With J.H. Donnelly, Jr. & J.L. Gibson. *Fundamentals of management* (8th ed.). Homewood, IL: Richard D. Irwin, Inc.
Human resource management: Foundations of personnel (5th ed.). Homewood, IL: Richard D. Irwin, Inc. (Adopted by over 240 institutions in 4th ed.).

1993

With M.T. Matteson. *Organizational behavior and management* (3rd ed.). Homewood, IL: Richard D. Irwin, Inc.
With M.T. Matteson. *Management and organizational behavior classics* (5th ed.). Homewood, IL: Richard D. Irwin, Inc.
With T.N. Duening. Internationalizing a business school: A partnership-development strategy. *Selections*, Autumn, 23-35.

1994

With P. Lorenzi, S. Skinner, & P. Crosby. *Management: Quality and competitiveness. Homewood, IL: Richard D. Irwin, Inc.*

With J.L. Gibson & J.H. Donnelly, Jr. Organizations: Behavior, structure, processes (8th ed.). Homewood, IL: Richard D. Irwin, Inc.

With E.S. Gardner, Jr. Productivity in the U.S. and Japan: A reexamination. *Interfaces*, November-December, 66-78.

With J.H. Donnelly, Jr. Management science and model building. In *Marketing manager's handbook* (pp. 356-369). Chicago: Dartnell Corp.

1995

With J.H. Donnelly, Jr. & J.L. Gibson. *Fundamentals of management* (9th ed.). Homewood, IL: Richard D. Irwin, Inc.

Human resource management: Foundations of personnel (6th ed.). Homewood, IL: Richard D. Irwin, Inc.

Job security: Job future ambiguity. *ILO Encyclopedia of Occupational Safety and Health* (in press).

Fred Luthans

A Common Man Travels "Back to the Future"

FRED LUTHANS

Conventional wisdom holds that writing an autobiography is not only difficult but can be insightful and sometimes even quite painful. I certainly have procrastinated about getting started on my story but perhaps not for the usual reasons. Frankly, my biggest concern has been who will want to read about such a "normal" life of a "common man."

In a nutshell, I was born and reared in Clinton, Iowa, a small Midwestern city, by loving, supportive parents and an older sister; went off to the University of Iowa for all of my higher education; got married to my college sweetheart, which has proved to be the wisest decision of my life; spent two years in the army teaching at the United States Military Academy; and then took my first and only academic position at the University of Nebraska. For the past 29 years I have taught, researched, consulted, traveled, and, most importantly, with my best friend and wife, Kay, reared four wonderful, never-in-trouble children, Kristin, Brett, Kyle, and Paige. Just as our parents were proud of our academic achievements, Kay and I are proud of the fact that all four of our children graduated from the University of Nebraska in four years with good grades, and all have gone on for further education. (Kristin has her master's degree, Bret and Kyle did both their master's and doctoral work at Nebraska and Paige took additional classes in photography.) Our pride extends to our son-in-law, Todd, and daughter-in-law, Dina, who are completing their master's. However, the spotlight has been shifted since Kristin and Todd

Management Laureates, Volume 4, pages 153-199.

ISBN: 1-55938-730-0

presented us with our first grandchild, Kourtney, surely the most beautiful baby girl, ever. Sounds pretty normal and common, if not dull, if love, support, family, education, and hard work are considered run-of-the-mill and boring.

To get an idea of style and format, I scanned previous volumes in this unique series put together by Art Bedeian. In contrast with what I thought I could contribute, I read with great interest as Fred Fiedler dramatically recalled that, as a young man he scampered up a gangplank of a ship taking him out of his Nazi occupied Austrian homeland while a couple of German storm troopers laughed mockingly. I was amazed and entertained by Bob House relating his "checkered" past growing up in what he termed "Unholy" Toledo. I was deeply moved by Larry Cummings revealing some painful events in his personal life and his decision to enter a Benedictine monastery. In other words, my colleagues' autobiographies seemed to range from historically interesting, to entertaining, to emotionally moving.

How can having your father be your hero, being married to the same wonderful woman for thirty-three years, and being thrilled by attending the Orange Bowl to witness Nebraska win the national football championship begin to compare to the others' autobiographies? Then it hit me. This "normal," "common man" story in these turbulent, uncommon times may indeed be quite unique, my niche, if you will, in the autobiographies.

Obviously, my ego is large enough (some have said too big), to think my story is worth telling. In my personal life, as well as my professional life, I feel I have been very lucky (external attribution?). However, I also feel I have created a lot of this "luck" through hard work (internal attribution?), perseverance, common sense, and, believe it or not, applying the behavioral/ reinforcement concepts I have espoused and written about over the past twenty-five years. These values and approaches to life have come from my personal life and professional career development. The following sections provide the highlights of this "common" man's travel back through his personal and professional lives that serve as the point of departure for his future.

GROWING UP IN CLINTON, IOWA

Both of my parents were from large families and German descent. (At early ages, all of my grandparents came to America from Northern Germany.) My father, Carl, was the youngest of 10 children (all now deceased) and the only one to graduate from high school. He worked very hard, long hours as the owner of a corner grocery store, with credit, free delivery, and plenty of personalized service to regular customers. Yet, he always found time to attend all of my activities and was very popular with my friends as the best pool player, card player, and ball player. He made a modest amount of money in his business and invested wisely in real estate and the stock market. Even though we became well off financially, we were deliberately not part of the country club set, as material things and "keeping up with the Joneses" never mattered to my parents or to me.

My father's true love was fishing and building things. He acquired the small farm my mother had grown up on about 15 miles south of Clinton right on the Mississippi River. With some help of a carpenter who owed my father money from the grocery store, my dad and I built a cabin along the river. I spent much of my youth on the river fishing, exploring (Huck Finn style) and water skiing (backwards, no hands, one ski, figure that one out, was my best trick). To this day, some of my fondest memories are centered around that cabin.

My father always vowed he would retire at age 45, and he did. Talk about early retirement! This turned out to be a wise decision because the year after he sold the store, the first supermarket was built in Clinton. Supermarkets were the demise of his type of mom-and-pop grocery store (population ecology?). He spent the rest of his life on his terms, mostly fishing the backwaters of the Mighty Miss.

My mother, Leona, came from a large family, all of whom are now deceased except for her sister, Florence, which leaves Florence and her husband, Van MacKenzie, as my only remaining aunt and uncle. Leona was as hard working as Carl and the "mom" of the mom-and-pop grocery store, as well as taking care of all of the household chores. My sister and my wife always claimed my mother waited on dad and me hand and foot, and, in retrospect, she certainly did. She was all anyone could ever ask for in a mother; her family was her life. All of my friends called her "Ma" Luthans. In addition to my parents, I have always received support, loyalty and love from my older sister, Nancy, even though she used to beat me up when we were kids.

Both my mother and father were supportive of anything I ever wanted to do, but it was clear they placed a priority on education. They never put direct pressure on my sister and me for grades and degrees, but they always gave us a dollar for each "A" we received. One of my most vivid memories involves an incident when Nancy was a senior and I was a freshman at the University of Iowa. Nancy received her certificate for making Phi Beta Kappa, and this was proudly displayed on our living room wall. Next to it, as a joke, my father hung a framed copy of a midterm delinquency report of the "D" I was receiving in one of my courses. Needless to say, I did not dare to end up with a "D" in the course!

I had a "normal" life in Clinton, an industrial river town of about 30,000. Life in the 1940s and 1950s revolved around sports, work, and school (in that order). In the fall we watched and played football, in the winter basketball, and in the spring track and field. In the summer we played baseball and tennis, went swimming and water skiing and played a little golf (on sand greens at the public 9-hole course). Starting at about age 12 I had every imaginable summer job. These included selling ice cream bars to businesses and factories, delivering papers, cutting lawns, working for my father at the grocery store, caddying at the country club (I caddied in the finals of the Iowa championship), selling shoes at JC Penney's, running an elevator at a clothing store, bailing hay, putting in fence and cutting weeds and trees at our farm, detasseling corn (sometimes for 16 hours a day), putting in a pipeline across the Mississippi (for $250 per week, such big money I almost didn't return to school in the fall), recreation leader at a park, cleaning out public schools, bellhopping and busboy at the local hotel and digging ditches for the gas company. I have always felt this diverse experience has given me an appreciation for and an understanding of manual and service work.

As a teenager, I attended Clinton High School, known as the River Kings to us and the River Rats to our opponents. My circle of friends and I were concerned only about sports and parties. I was always a fast runner, which carried me to a degree in team sports, but I never did excel in football or basketball (rode the bench). However, in track I was the best hurdler on the team and one of the top in the state. My senior year I was captain of the track team, King of the Clinton Gateway Classic Track Meet, and recruited by the University of Iowa and a few other smaller schools.

Intellectual curiosity and pursuits were absent. I would usually take books home at night but never looked at them. I was of the opinion that only the nerds did homework. I was considered a good, but not outstanding, student. Some of my high school teachers were later somewhat surprised to hear that I had received a Ph.D. and was a college professor.

A PERENNIAL STUDENT AT THE UNIVERSITY OF IOWA

I never remember having a choice or even considering going anywhere except the University of Iowa. My sister was already at Iowa, and everyone in Eastern Iowa was oriented toward the University in Iowa City. My undergraduate years (1957-1961) were basically an intensified extension of my high school years. Because I was a decent math student, it followed that I would become an engineer. However, once again social and sports activities prevailed.

My freshman year I pledged Phi Delta Theta social fraternity, which turned out to be very similar to the popular movie, "Animal House." Even though my grade-point average was only about a C+, it was the highest in my pledge class, and all of my best friends flunked out after the first year. During my sophomore year I met and started to steadily date my wife-to-be, Kay. In the meantime, I found out that I was not cut out to be an engineer, but by that time I had so many hours in math that it was easiest for me to become a math major in liberal arts. In my senior year, I also began to take some business classes because I knew I did not want be a mathematician or an actuary.

Although I was on the track team, I never took it very seriously. I was a mediocre college hurdler, at best, and pole vaulted when they needed a body in dual meets. My claim to fame was that I was a member of our hurdle shuttle relay team that would place at big meets, such as the Drake Relays, and one time we appeared on national TV on the Wide World of Sports. I also like to tell my kids and students the story of how I broke a world record.

At the end of my senior season, some of our team members entered the Midwest AAU meet at the University of Chicago. I broke the existing world record in the high hurdle event. Now, as I ask my amazed students, do you doubting readers want the rest of the story (á la radio personality Paul Harvey where he gives some historical facts and then provides "the rest of the story" after a commercial break)? To generate some interest, the promoters of this track meet inserted a gimmick event, the 100 yard high hurdles (those of you who know track know that 120 yards, and now 110 meters, is the standard distance). However, at that time the 100 yard distance was still in the record book but had not been run for many years. I came in under the official world record time that dated back to the 1930s. I only got third in this race, so I did not hold the odd-ball record, but I did break the world record. I have the newspaper clipping to prove it. So there! As my students chuckle and shake their heads, I come back with, "Well, how many of you have broken a world's record?" One time a guy in the back raised his hand, and I recognized him as a star swimmer who had broken the 50-yard freestyle record that year. I came back with, "Okay, err, besides you Peter, who else has broken a world record?"

So much for my athletic career. I really did enjoy the experiences, which only former athletes can appreciate. I was able to get to know, as teammates, former Olympians, such as Ted Wheeler (who later became the track coach at Iowa) and Charles "Deacon" Jones, and interact with other well known Iowa athletes, such as Don Nelson, the long-time player for the Boston Celtics and currently of NBA coaching fame. Even today, Tom Osborne, the highly successful and respected Nebraska football coach, not only knows me as a professor, neighbor and friend (I have helped him recruit over the years), but also as a former Iowa hurdler.

After graduation, again with my parents' support, I decided to enter the new MBA program at Iowa to enhance my job opportunities. An academic career had not even crossed my mind. Initially, I started graduate school with the same approach I had taken as an undergraduate—socializing first, studying second. I shared an apartment with former fraternity brothers above the infamous Whiteway grocery store in downtown Iowa City (Alex Karras, the famous pro football player and later commentator/actor was a legend for his exploits in these same apartments). Then, at the beginning of the second semester in February 1962, something happened that changed my life.

This life changing event started with a typical Friday night party at the apartment. Kay, by the way, was now in school in Northern Colorado. I was taking strong aspirin for the "dry socket" resulting from my wisdom teeth being removed and made the mistake of drinking a lot of rum. The morning after I scurried to the bathroom to vomit. This was not unusual for me because I always had a weak stomach. But this was red vomit, and I had drunk white rum. I was vomiting pure blood. I remember thinking to myself, "Oh-oh, this is one of the danger signals of a serious problem." I unsuccessfully tried to rouse my roommates to take me to student health, but they were still sleeping off the night before. I then ventured out into the cold winter day to drive myself. About half way to my car parked several blocks away, I collapsed into a snow bank, in shock from the loss of blood.

Lying in the snow bank was a moment of truth, a turning point in my life. Here I was, 22 years old, in the prime of life. I had everything going for me, and things had come relatively easy for me, but I had no direction, no burning desire of what I wanted to do, and now I was in serious trouble for the first time.

A passing student helped me to my car and drove me to the big, socialized medicine labyrinth called the University of Iowa Hospital. I was put on a gurney in the hallway and forgotten. Through the fogged state I was in, I remember grabbing a "Candy Striper" passing by and asking her if she could get someone to take a look at me. A nurse then showed up and tried to take my pulse. The following conversation is burned in my memory. She said, "I am having trouble getting a pulse on you." I said, "What does that mean?" She said, "That you are about gone." At that point she got things moving "stat." I had a plastic tube stuck down my nose into my stomach to pump out the blood and an IV stuck into my arm to start pumping in some new blood.

To make a short story even longer, I had a bleeding ulcer and had lost about half the blood in my body. The bleeding thankfully stopped without surgery. Transfusions, medication, and a strict diet took care of the problem. I had considerable weight loss on an already slim body and for the first time felt very scared and vulnerable. My parents got me off the big ward and put me in a private room with the head of internal medicine, Dr. Hamilton, as my doctor. After spending a couple of weeks in the hospital recuperating, Dr. Hamilton gave me a stern lecture. He told me to move out of the apartment, get a quiet room, eat regularly, and in essence, get my act together. I religiously followed his instructions and for the next several years, I not only never had an alcoholic drink or a beer, I never even had a soft drink (or any carbonated beverage). More importantly, for the first time I became a serious student and enjoyed it.

I discovered the academic world by taking challenging, well-taught courses in the remainder of my MBA program. In particular, I remember my labor (Chester Morgan) and macro (Anthony Costantino) economics courses and the dynamic teaching style and ideas of Harvey Bunke, who later became dean at Indiana University and long-time editor of *Business Horizons*. However, the course and professor that had the greatest impact on me was Organization and Management Theory taught by Henry H. Albers.

I consider Albers, a Yale educated labor economist, to be one of the true pioneers in the history of modern management thought and my mentor. Although preceded by the well-known texts by Harold Koontz and Cyril O'Donnell and by George Terry, Albers had one of the first principles of management texts, published by Wiley in 1961. When I was an undergraduate, we used excerpts from his upcoming text, and I received a "C" in the course. In Albers' MBA class, however, now that I was a serious student, I received an "A" (and in all subsequent graduate courses). I was greatly stimulated by Albers' classroom ideas and writings and truly became interested in and saw the potential for the development of the field of management. I finally had a goal and something I wanted to vigorously pursue. I wanted to make a contribution to this emerging field.

Albers was a thinker and a doer. I still consider him to be one of the most intellectually stimulating people I have ever known or been associated with. In this MBA course, he not only introduced me to the wonderful world of Chester Barnard and Herbert Simon but also to his own dynamic process. The functions of decision making, communication, human performance and feedback control served as the conceptual framework for his book. I still use his framework today as part of my own MBA classes. His ideas were not restricted to the management field, but also included history, philosophy and, especially, politics. In addition, Albers had an entrepreneurial flair for venturing into new ground, such as nontraditional education. He encouraged me to go on for my Ph.D. under his direction, and I didn't ever consider saying, "No."

In the fall of 1962 I began my Ph.D. program, and the mainstream management field was just starting to emerge. I still feel extremely fortunate to have had the program of courses and major professors that I had at that time. Albers, of course, was my chair and mentor, but I also had Max Wortman, then a brand new assistant professor out of the University of Minnesota, for an area in personnel/labor relations; Cal Hoyt, an industrial sociologist out of Cal Berkeley, for an area in organization theory; and Milton Rosenbaum, a social psychology professor from the very strong Iowa psychology department. This grounding I had in management and organization theory, human resources (HR), and psychology was quite unique for the times, especially the blending of these particular areas. Obviously, each of these three areas by itself was going strong in various schools around the country, and even personnel/industrial relations and psychology were blended together in strong programs at the Industrial Relations Institutes at schools such as Minnesota and Illinois, and the same for the Industrial Psychology programs at Cal Berkeley and Yale. However, the general management and macro organization theory component, which I also had, was largely missing from those other programs. This broad Ph.D. program at Iowa with these outstanding professors gave me a distinct advantage in later developing my approach to organizational behavior.

I greatly enjoyed my doctoral student days at Iowa. My closest student colleagues were Bill Reif and Jim Walker. Although a year behind me, Bill and I were very close because we were both members of the track team and fraternity brothers as undergraduates. He became a long-time professor at Arizona State and served as Associate Dean and Acting Dean. He subsequently joined Jim's consulting firm in Phoenix. Jim was in the vanguard of the HR field with his book on strategic human resource planning as well as being a successful full-time consultant.

Kay and I got married my second year in the program. We lived in a small, two room apartment literally in Herbert Hoover's boyhood backyard in West Branch, a small town right outside of Iowa City. Kay taught English at the high school, and I wish I had her teaching evaluations! She got the attention of the worst, as well as the best, students. They all greatly respected her. She made both Shakespeare and grammar understandable and fun. Tough, but fair, she once critiqued one of my lectures and gave me a C+. Kay's wonderful parents, William and Mildred Meldahl, were very supportive of me. I immediately became part of their family, which included Kay's older brothers Bill and Roger and their families. Now retired, Bill was a top-level executive with Westinghouse and Roger was a vice president of marketing for the Union Pacific. I have used them as a real world sounding board for some of my ideas through the years.

My Ph.D. program went smoothly. My dissertation analyzed the centralized control (from the vice president of academic affairs level) of decentralized faculty promotions (from the dean and department head levels). I used a large sample of universities with colleges of business. I basically found that there really was no centralized control of faculty promotions. This finding had implications for building quality and consistency across these universities. I suspect that this still largely holds true today. The dissertation was published in book form by the University of Iowa Bureau of Business Research, and an article from this study was published in the *Academy of Management Journal.* Before I graduated, I had several published articles co-authored with Max Wortman. Max taught me how to write articles, and with him I co-authored my first book, titled *Emerging Concepts in Management* (Macmillan) in 1969.

In 1965, at age 25, I graduated with my Ph.D. This is one reason why those in the field who meet me for the first time are often surprised at how relatively young I am (which of course I now love to hear). For my age, I have been around the academic field of management for a long time. As an undergraduate at Iowa I had received my officer's commission through the Army ROTC and had received educational delays while in graduate school. Once again, I had no choice of where or what I would do with my life. The day after graduation in May of 1965, I flew to Georgia and reported to Infantry Officers' Basic Course at Ft. Benning. I knew right away my Ph.D. didn't mean much to the big, burly sergeant who yelled at me to hit the dirt and give him ten. "Welcome to the U.S. Army, Doc!"

WEST POINT DAYS

If I had left my fate up to the U.S. Army, I would have undoubtedly been assigned as a platoon leader in an infantry unit in Viet Nam. The time was 1965, and this was the beginning of the big build-up. The Army, of course, made use of M.D.'s being inducted but made no provisions for Ph.D.'s. With the help of Colonel Holmes, the head of the ROTC unit at Iowa and a Military Academy graduate, I took the initiative to be assigned to the United States Military Academy, commonly known as West Point. The Military Psychology and Leadership Department liked my credentials for teaching psychology and leadership. They arranged for my duty assignment, but felt I needed a little polish first, so I spent nine weeks in infantry training.

All of my fellow trainees had been assigned to Viet Nam, but I had orders for what my training sergeants referred to as "Hudson High" or the "Ring Knockers School." At that time the protests and criticisms of the war were not very visible. All my army buddies were enthusiastic about getting over to "Nam." I was happy about my assignment, not so much to avoid the war, but because I really was eager to begin my professional, academic career. Another factor was confidence. I was, at best, only an average infantry officer and felt very ill-prepared, even relative to my fellow trainees, to lead an infantry unit in Viet Nam. I felt very well prepared to teach cadets at the Academy. Fortunately for me, and I might add for those I would have had under my command, I had the opportunity to teach at the Academy.

Ft. Benning was the real army. West Point was a very unique and, at least for faculty members, a wonderful place. It was full of historical tradition and, for me, truly lived up to its slogan of "Duty, Honor, Country." My family, especially my dad, was very proud that I was teaching at West Point. The department of Military Psychology and Leadership (MP & L) was relatively recent, and its founding was attributed to the efforts of Dwight Eisenhower. Other organizational behavior (OB) academics, such as Dan Ilgen and some of Fred Fiedler's doctoral students, later served in this same department. MP & L was unique within the Academy because it was organized under both the academic branch and the military side, the Department of Tactics. Thus, I was able to belong to and be an active participant of both very distinctive sides of the Academy. Unlike those assigned to the other academic departments, I was involved in at least observing the military side of cadet life at West Point.

From "Beast Barracks," when the new cadets had their heads shaved and were put through eight weeks of hell, to Camp Buckner, which was attached to the post and was where the cadets received military training in tactics and weapons, to outside class activities for the cadets, such as formations, marching on the famous trophy point plain in full dress uniform, and walking off demerits in the Central Area, I was there.

Officers who taught only in the academic departments did not really have such exposure to these "fun and games." Also, about once every couple of months, I would draw the assignment of officer in charge or OC. The OC was in charge, the commandant's representative, of the entire Academy and all the cadets after regular duty hours, from 5:00 p.m. till 7:00 a.m. Wearing the big OC badge on the shoulder gave one a considerable sense of power but also a sense of fear of the responsibility. When I would make my inspection rounds in the cadet barracks, the word would spread very quickly that the OC was in the area.

Cadet life was very ordered, and it was commonly understood that there was a right way to do things, a wrong way, and the West Point way. Regulations for the rooms included: "All books should be pushed to the back of the shelves and not be placed tangent to the edges" and "The shelves in the medicine cabinet should have hairbrush, soap dish, and razor on top; toothbrush and toothpaste in the middle; shaving cream, deodorant, and water glass at the bottom." I had no clue about all these rules (their Tactical Officers did), but as OC I would still randomly barge into some of the rooms for inspections. The cadets present would drop what they were doing and smartly brace up against the wall. Wow, what a sense of power and authority. What a contrast with my students through the years at the University of Nebraska!

However, remember this particular OC was a guy who wasn't even sure how to put on his brass properly, let alone having it shiny and accompanied by spit polished shoes. In fact, during a break in my psychology class one day, a fellow officer called me aside and whispered that my brass was on upside down. Flustered and embarrassed, I thought, "How am I going to get out of this one?" because I was sure some of the cadets in my class noticed, too. So, I went back into class and said to the cadets, "Okay, how many of you caught the perception experiment I was conducting on you?" (The honor code didn't apply to me, did it?) By the way, about half raised their hands.

On another occasion, I was dressed in uniform, as usual, and was hurriedly leaving home for work. I reached up into the closet and put on my hat. As I was walking from the parking area to my building, I came upon a senior officer and smartly threw him a salute. He did a double take and gruffly said, "What the hell do you have on your head?" I quickly felt up there and pulled it down. You guessed it, I had put on my ragged, old plaid hat I had worn as a graduate student. The spit and polish world of West Point had never seen anything quite like this. However, I did take solace in the legend of when Edgar

Allan Poe was a cadet. He supposedly showed up at a full dress parade wearing only his breast straps and sword, nothing else.

The biggest fear I had of being OC (besides being discovered for the disheveled rookie that I was), was the responsibility of being in charge of 4,000 potentially rowdy young men. I will never forget one OC stint sitting on the "poop deck" where MacArthur had given his famous "Duty, Honor, Country" speech. I was having my supper overlooking the 4,000 cadets below "chowing down." This particular evening was during the few days remaining in the countdown to "The Game," with Navy, and the troops below were very restless and particularly animated. The cadets were known to do some bizarre things before the game, such as stacking the dining hall furniture pyramid fashion and ordering plebes to climb the mountain. While sitting there having this wonderful meal (the cadets ate very well) served by several enlisted men, I had visions of a James Cagney movie prison riot on my hands below. Then I remembered my management principles. I called the cadet officers in the chain of command over to my table and said, in my best OC voice, "Gentlemen, I am delegating to you the control of any problems that may erupt below." In my OB voice I quickly added, "I know I can trust you to handle this and that you will do a good job." They quickly fanned out and things immediately calmed down and there were no problems. I knew my management and behavioral education would do me some good.

As you can tell, some of my fun memories from my days at the Point were from the military side. Frankly, both Ft. Benning and the military side of the Academy were a good break from the graduate school grind. It put me, not in the real world, but in another world that had therapeutic value and motivated me all the more for an academic life. In other words, at least for me, I could not have asked for a better place to start my academic career.

At that time, all West Point faculty members were uniformed, active military officers, and there were no female cadets. Most of the faculty had just come out of master's programs relevant to their teaching assignments, but there was also a core of permanently assigned faculty members with excellent academic credentials. Almost all were field grade officers (major, lieutenant colonel or colonel), with the exception of a couple of lieutenants, such as myself. Even though I was always outranked, I had considerable status among both faculty and cadets. This was because they knew that as a lieutenant (the second year I was promoted to captain), I had to have a Ph.D. Having just come out of master's programs, the high ranking officers greatly respected my advanced degree. Everyone treated me great, both personally and professionally. I have very fond memories of my fellow officers and the cadets.

Unlike new Ph.D.'s taking their first jobs in universities, I was under no pressure to hit the ground running as far as new class preparations and getting out publications. I was initially assigned to help research the validity of what was called the Aptitude for Service ranking. This was essentially a forced peer rating where each cadet ranked from top to bottom everyone in his company (about $N = 60$) according to the criterion of how he would view this fellow cadet as an effective leader in a combat situation. We found this rating to have the best predictive validity for a number of success criteria, such as those who made the outstanding/early promotion lists, retention in the service, and even those who received combat medals and made general. This peer rating predicted better than academic rank in class, involvement in sports, or grades in tactics courses and ratings by their TAC (tactics) officers.

Where one came out on the peer rating was primarily used to determine the cadet chain of command. The first captain had the highest rating. However, those who were ranked at the bottom were brought before an Aptitude Board (to which I was occasionally assigned) and were sometimes dismissed from the Academy. Because these "aptitude cases" sometimes stood very high academically, we had to show documented research evidence of the validity of the peer rating to upset parents and their senators whose appointees had been dismissed.

Although I was involved in this and other leadership research projects throughout my two years at the Academy, I also taught the required basic psychology course to Yearlings (sophomores) and a course called Leadership to Firsties (seniors). In typical Army fashion, we had lesson plans and much planning and coordination, which I found quite beneficial in launching my teaching career. Although I had taught statistics and principles of management as a graduate student at Iowa, I found this structure and support had lasting value to my teaching.

At least at that time, education at West Point was based on the Sylvanus Thayer model (the most famous West Point educator of the early 1800s who has the hotel and the main academic building in which I taught named after him). The distinguishing features of this educational system included small classes, much discussion and, most importantly, a short quiz (or what was called a "writ") at the conclusion of each class period. In retrospect, this approach was very effective, especially for this type of faculty and students. Although the cadets would snap to attention as I entered the classroom, at the "at ease," they would quickly transform into typical college students. The Thayer system made sure that they were talkative and very well prepared.

I found the famous (or perhaps in more recent years the infamous) honor system to be very real and to have a definite positive impact on the culture of the Academy. The meaning of the honor system was demonstrated in class one day as we were discussing the overwhelming evidence on the relationship between smoking and lung cancer as an example of making an inference of causation from correlation. I had a star track athlete in my class, and I made the offhand remark, "You never smoke, do you, Mister." He looked around quickly and sheepishly replied, "Yes sir, I do." I had no intention of putting him on the spot like that, but he had to tell me the truth. Now he was in trouble with his coach and teammates for smoking, but he always had to tell the truth. Also, if one of his friends knew he smoked, the friend would be obligated to step forward if the track star had not told the truth. The honor system also meant that there were no locks in the gym locker room or the barracks. On their first assignments in the real world, West Point graduates invariably had their wallets stolen because they were so used to the West Point honesty culture.

I found the cadets to not only have good character but also to be very bright. Although they had extremely rigorous and structured lives at the Point, once past the facade, most of them had a lot of fun and did not take themselves or the system too seriously. I spent some time at both the Naval and Air Force Academies on exchange visits and thus also got to know their programs fairly well. My quick reaction in comparing the three would be that West Point had the historic tradition of producing famous generals, such as Lee, Grant, MacArthur, Patton, and Eisenhower; the Air Force Academy had the newest facilities and the most up-to-date for that time academic program; and Annapolis was kind of the "neatest" place to be. I frequently have been asked to evaluate the education received at the service academies and whether I think they should be allowed to continue. My reply is always the same. They are very effective for their intended purpose: to educate and train career military officers.

Sports reigned supreme at West Point, which fit in nicely with my own background. The staff and faculty had a very competitive basketball league. Every long lunch hour was spent playing basketball. I also played a lot of tennis, learned to play squash and, in the winter, we even had our own ski run on the post. The cadets had a full program of intercollegiate athletics including all the normal sports, but also hockey, boxing, soccer, rugby, lacrosse, and even a 150 pound and below football team. The football and basketball teams were at the beginning of their slide from the big time. Paul Dietzel, who had won a national championship for Louisiana State University (LSU), was hired to bring the football team back to glory, and Bobby Knight was the basketball coach. You can imagine what a highly disciplined defense and patient offense Knight ran at West Point. The point guard, who was Knight's floor general, was Mike Krzyzewski, who, of course, later coached national championship teams at Duke. While I was there, the basketball team was relatively successful and propelled Knight into the big time at Indiana University, but the football program never did get turned around. Obviously, the service academies have a lot to overcome in recruiting the type of athletes needed to compete with top teams, such as the "Big Red Machine" at Nebraska.

West Point turned out not just to be the fulfillment of my service obligation, but very beneficial to my professional development. Not only did I not have the pressures facing new assistant professors, I learned a lot. First of all, I had the time to read and do basic research in a learning sense, rather than just for publication. I had the chance to actually teach psychology and this, of course, meant that I really learned the material. The leadership course I taught was from a social psychology perspective and this again gave me a knowledge base and confidence that I did not receive as a grad student. In other words, I was able to get a degree of depth in psychology from study, research and teaching that my graduate management program, with only a minor in psychology, did not permit.

The real topper for me professionally at West Point, however, was that I was allowed to enroll in courses as a post-doctoral student in the Columbia University Graduate School of Business. I took seminars from Bill Newman, who was formulating some of the early ideas on strategy, and Leonard Sayles, who had sort of an anthropological perspective on the early development of organizational behavior and leadership. I also was allowed to sit in on the Executive Development Programs run by Columbia at the Harriman estate called Arden House in the Hudson River Valley. Although most of the major early contributors to organizational behavior taught in this program, Chris Argyris had the biggest influence on me. To this day, I give Chris a lot of credit in formulating my ideas and approach to the field.

Besides the significant contribution West Point made to my professional development, Kay and I had a great time socially and culturally. The highpoint was when our first child, Kristin, was born at the West Point hospital. We received a whopping bill for $8.50 (for Kay's meals), and we always like to tease Kristin that "You get what you pay for!" The year or so before she was born we took great advantage of the cultural activities of New York City. Through the United Service Organization (USO) we received free tickets to attend all the Broadway shows and some concerts at Carnegie Hall. Kay had directed the school plays in West Branch and we really got into the New York theater scene—free of charge.

We also were completely accepted into the West Point social circle. The wives of high ranking officers really knew how to entertain and give dinner parties. In addition, there were many formal social affairs, such as the Commandant's and Superintendent's Receptions and free concerts at West Point put on by celebrities, such as Sammy Davis, Jr. It was also neat to have colleagues in other departments, such as Pete Dawkins, the well known Heisman Trophy Winner and living legend in the Army, who taught in the social science department, and Norman Schwarzkopf, whom I did not know, but who taught engineering mechanics at the same time I was teaching leadership to the cadets. Something was wrong here, but as I now jokingly tell my students when I name drop, luckily my door was open so that "Stormin" Norman could overhear my lectures on leadership that proved useful to him during the Gulf War.

The good times were very good at West Point. We had a lot of fun. Unfortunately, there were also some very sad times. Some of the fine young men who had been in my classes as seniors and then went off to Viet Nam, were sent back to be buried at the West Point cemetery. The tragedies surrounding the Viet Nam War and its aftermath for the Class of 1966, which included some of my students, were depicted in a best-selling book published in 1989 called *The Long Gray Line*. This 1966 class eventually had more dead (30) and wounded (over 100) than any other West Point class. As their classmate, turned author, Rick Atkinson later noted in the book, "The survivors came home to heckles and, in a few instances, spittle." I vividly remember walking between classes and having very young kids who had been bused into West Point for a field trip for the day, shouting and giving obscene gestures out the window at me and, of course, I would never dare wear my uniform when I attended my classes at Columbia or went into the city. These were not good times for either "side" of this war.

When my obligation was up, I was offered a permanent appointment at West Point. The two years were a great experience, but I knew this was not something I wanted to do for the rest of my life. I entered the academic job market in the spring of 1967. Because I had the connections at Columbia, I could have taken a tenure track job there, but I soon figured that parking alone would take up a significant portion of my salary offer of $9,200. Kay and I wanted to rear our family in the Midwest, so I interviewed and received offers from the Universities of Michigan, Missouri, and Nebraska. The Michigan job was for an assistant professor position at $11,200, and I would have enjoyed working with Ed Miller, my main contact there who is still a good friend today. Both Missouri and Nebraska offered to bring me in as an associate professor. I liked both but settled on the Nebraska position for $11,700 (Missouri offered me $12,500). Richard Hodgetts, my closest colleague and co-author through the years, had preceded me at Nebraska by one year. I will never forget when I asked him about Nebraska on my recruiting visit. He bluntly said, "I wouldn't come here if I were you." I have been diligently following his sage advice ever since.

EARLY YEARS AT NEBRASKA

Very unusual in the academic world, I have spent my entire career at the University of Nebraska. I was hired in 1967 to teach the first management courses. Personnel and industrial (later called production and now operations) management courses were being taught, but general management courses at both the undergraduate and graduate levels were just beginning to be offered. A human relations course (using Keith Davis's popular text) was being taught but, of course, no organizational behavior. I taught personnel administration and principles of management to undergraduates and developed a new course in organization and management theory for grad students. The very best student in my first classes was an airforce vet by the name of Dick Daft. He took a job in Marshalltown, Iowa, Kay's hometown, and when we got together there socially a few times, I finally convinced him to go on to graduate school. Many years later, I had the pleasure of successfully nominating Dick to become a Fellow of the Academy of Management.

My first year at Nebraska we had only two departments in the college, business administration and economics. Then, before my second year, business administration was split into departments of accounting, finance, marketing and management. Upon my suggestion and strong recommendation, we interviewed and hired, as the first chair of the department of management, my mentor from the University of Iowa, Henry Albers. From the get-go, Albers starting building an excellent department. He started a management-oriented master's degree at Offutt Airbase in Omaha and landed a large government contract to do management training in Micronesia. He also developed management development programs, which we gave throughout the state, and implemented the latest curriculum for our students. Most importantly for me, he created the expectations and provided the support for my research and writing.

Upon my arrival at Nebraska, I immediately hit it off with my slightly younger colleague, Richard Hodgetts. He was, and still is, full of energy and ideas. After my first book with Max Wortman in 1969, which was mainly readings, Hodgetts and I collaborated on our first book together, *Social Issues in Business* (Macmillian, 1972). My work with Hodgetts has proved to be a very productive, satisfying and mutually rewarding relationship to this day. Dick and I have become as close as family, and I was honored to serve as his best man when he married Sally in Lincoln in 1970. After hearing all the horror stories between co-authors over the years, I feel very fortunate to have such a great writing partner and pal as Richard Hodgetts.

Besides having Albers as my department chair/mentor and Hodgetts as my colleague, I was able to get immediately involved with some outstanding doctoral students. Jim Francis (long-time chair and now professor at Colorado State), Don White (long-time chair and now professor at Arkansas), Bob Ottemann (long-time professor and current chair at Nebraska-Omaha), Jerry Wallin (long-time professor at Louisiana State University), and Bob Kreitner (long-time professor at Arizona State) come quickly to mind. Although I never had Francis in my classes, we worked on some research projects together while he was finishing up at Nebraska, and he has been a valued colleague and friend over the years. My first doctoral students, White, Ottemann, Wallin and, especially, Kreitner, all played a vital role in my development of organizational behavior modification (O.B. Mod.).

The seeds for O.B. Mod. go back to the late 1960s when I was doing management-by-objectives (MBO) work for the Nebraska State Mental Health System under the head administrator, Dr. Robert Osborne. After giving MBO training and consulting around the state, Dr. Osborne, a psychiatrist with administrative interests and effective skills, challenged me to start thinking about and using behavior modification approaches and techniques in my research and training work. He had found that B-Mod was having some spectacular results on the clinical/treatment side of mental health and wondered why I didn't apply some of this approach to human resource management.

For example, a patient in the Lincoln Mental Health facility had not uttered a sound for many years. Conventional treatment techniques from the medical and psychotherapy models had no impact. Then, Dr. Osborne allowed an educational psychology professor steeped in B-Mod to take over the case. This behavioral technologist noted that the patient, because of his seniority, had one of the best rooms overlooking a wooded area, the best seat in the TV room always staked out, and full privileges in the commissary and grounds. In other word, this patient was being reinforced for his non-communicative behavior. The B-Mod expert eliminated these contingent reinforcers (not punish) and put the patient on a token system. Now the patient had to earn benefits and privileges by exhibiting desirable behaviors of first uttering sounds, then words, then talking. In a matter of weeks he was talking. The patient's problems and trauma were not cured, but now at least the other techniques, such as psychotherapy, had a chance. The psychiatrists were resistant to these techniques, but Dr. Osborne was open to new ideas and approaches and thought they should also be tried on the administrative side as well. I quickly agreed with him.

Reaching back to my Iowa psychology days, I said to myself, why don't we apply operant-based learning approaches and techniques to human resource management. It was one of those "eureka" things that just kind of hit me all at once. I tried to find something in the literature and came up blank (we had no computerized searches in those days). So, with Don White (my very first doctoral student), we wrote what I thought, at the time, was the first operant-based approach to human resources management article titled, "Behavior Modification: Application to Manpower Management" for *Personnel Administration* which came out in the summer of 1971. Later, I came across articles by Owen Aldis, "Of Pigeons and Men" published in the *Harvard Business Review* ten years before, the now classic piece by Walter Nord titled "Beyond the Teaching Machine: The Neglected Area of Operant Conditioning in the Theory and Practice of Management" published in *Organizational Behavior and Human Performance* in 1969, and the results of some research on "The Application of Behavioral Conditioning Procedures to the Problems of Quality Control" by Everette Adam and William E. Scott, Jr. reported in the *Academy of Management Journal*, the same time as our article came out.

I took Dr. Osborne's advice and devoted my research/writing and training work in the late 1960s and early 1970s to an operant (Skinnerian) based approach to management of human resources. The research program started with testing an O.B. Mod. training intervention in a light manufacturing firm in Omaha. This became the dissertation of my second doctoral student, Bob Ottemann. Along with considerable practical help from David Lyman, a student of mine who was getting his doctorate in educational psychology with direct experience using behavior modification techniques with parent groups, we trained a group of production supervisors. This training basically amounted to using the behavioral approach that we had been writing about to improve the performance of work groups. To test the model, Bob Ottemann and I used a pretest-posttest control group experimental design. The O.B. Mod. training intervention with nine supervisors in an Omaha manufacturing plant was the independent variable, and their combined units' engineering measured productivity was the dependent variable. The supervisors trained in the use of O.B. Mod. significantly increased the performance of their workers, and the matched control group of supervisors, at the same plant, who did not use O.B. Mod. had no increase in the performance of their workers. We reported these positive results at the 1975 Academy of Management meeting. This experiment gave me some confidence that we had something here, and stimulated me to do a stream of research over the next several years in all types of field settings to test the external validity of O.B. Mod. as an approach to improve employee performance.

Simultaneous to the basic field research on O.B. Mod., my doctoral students and I also continued writing practitioner-oriented articles on this new behavioral approach. Bob Kreitner and I had originally come up with the term organizational behavior modification or simply O.B. Mod., and I used it for the first time in the initial edition of *Organizational Behavior* (McGraw-Hill, 1973). Articles with Kreitner (1973, 1974), Lyman (1973), and Ottemann (1973) at about the same time expanded and spelled out the theoretical perspective and O.B. Mod. implementation steps. This culminated in the book I co-authored with Bob Kreitner titled *Organizational Behavior Modification* (Scott, Foresman, 1975).

Although Bob worked closely with me in his doctoral program, I steered him over to the operant learning and behavior modification courses that were very much in vogue at the time in educational and counseling psychology. Bob was, and is, a very hard working and talented synthesizer and writer, as his currently successful textbooks attest. Right after Bob received his Ph.D., we drew from his extensive knowledge of operant learning and, by then, my own considerable research, writing and training experience on O.B. Mod., to write what I still consider my major contribution, the 1975 O.B. Mod. book.

The O.B. Mod. book provided the historical and theoretical foundation for our five step application model of: (1) *identify* the critical performance-related behavior; (2) *measure* the baseline frequency of this behavior; (3) *analyze* the antecedents and especially the consequences (the A-B-C functional analysis); (4) *intervene* with positive reinforcers to accelerate the functional performance behaviors and put the dysfunctional behaviors on extinction and, as a last resort, apply punishers closely followed by reinforcers on the desirable alternative behaviors; and (5) *evaluate* to insure that performance is in fact improving, and if not, go through the steps again.

My doctoral students in the 1970s all worked on various aspects of O.B. Mod. Ken Bond and Pete Van Ness did dissertation studies on the methodologies for testing O.B. Mod. Doug Baker (long-time professor at Washington State), Tim Davis (long-time professor at Cleveland State), Terry Maris (currently Dean at Northern Ohio), Mark Martinko (long-time professor at Florida State), Bill Ruud (currently Dean at Boise State), Jason Schweizer (currently at Thunderbird), Charles Snyder (long-time professor at Auburn), and Ken Thompson (long-time chair and now professor at DePaul) either did dissertations or co-authored articles with me reporting research or further development of O.B. Mod. For example, Baker, along with my long-time colleague Bob Paul at Kansas State, helped on perhaps the strongest methodologically and most comprehensive study on the impact of O.B. Mod. on salespersons' performance behaviors, which was published in the *Journal of Applied Psychology*. I also collaborated on successful O.B. Mod. studies reported in articles with Maris, who did a reversal design in a bank, Schweizer, who did a multiple baseline design in an entire small manufacturing plant, Snyder, who analyzed the impact on productivity measures in a large hospital, and Thompson, who analyzed the impact on organization development and organizational culture.

Besides Kreitner, Mark Martinko and Tim Davis made the largest contributions into my development of O.B. Mod. in the 1970s. I first met Martinko while doing O.B. Mod. work with Western Electric in Omaha. He was a company trainer assigned to assist us. I talked him into working with me in a Ph.D. program in the mid-1970s, and I have been working with him on articles and books ever since. Mark, along with his contemporary at Nebraska, Ken Thompson, have been great colleagues and friends of mine over the years.

Tim Davis came into our Ph.D. program via England and a strong master's program in interactive communication from Nebraska. I had many arguments, if not battles, with Tim over theoretical perspectives and methodology. I give him credit for getting me off the radical behaviorism plateau I was then on and helping me expand into a more social learning paradigm. Although I still remain more of a Skinnerian behaviorist than a Banduraian social learning theorist, Tim and I went beyond a strict operant-based approach in a series of articles incorporating a social learning approach. These were published in the late 1970s and early 1980s in *Academy of Management Review, Journal of Applied Behavioral Science*, and *Organizational Dynamics*. I now felt so strongly about the importance of social learning theory in O.B. Mod., that I convinced Bob Kreitner we needed to update our 1975 book. We did this in our "Radical Behaviorists Mellowing Out" article for *Organizational Dynamics* and the ten-year revision of our original book that we titled *Organizational Behavior Modification and Beyond* (Scott, Foresman, 1985).

The social learning theory expansion of O.B. Mod. basically recognized both overt and covert antecedents and consequences, a four term Situation-Organism-Behavior-Consequence (S-O-B-C) contingency framework that recognized the existence of cognitive mediating processes (O) between the antecedent stimulus and the behavior, and the important role of modeling and self-control processes. This social learning expansion greatly enriched the theoretical foundation and explanation and thus helped gain acceptance of O.B. Mod. by the organizational behavior field. However, like the ever popular more cognitively-based OB theories, the social learning version of O.B. Mod. has never, in my opinion, held up to the more predictive and control/management power of the earlier radical behaviorism approach. Although Tim Davis did some empirically-based studies on self-management from a social learning base, we do not have the stream of research support that I feel we do have for the more parsimonious operant-based approach to O.B. Mod. To this day, O.B. Mod. still plays a major, but certainly not only, role in my teaching, research and consulting/training.

Although I was already very busy with the O.B. Mod. work and writing the social issues text and more general management articles with Hodgetts, Albers began prodding me immediately after he had arrived in Nebraska in 1968 to write a mainline text on my emerging field of organizational behavior. When I first began writing *Organizational Behavior* in 1968, I was not aware of any other such text. Keith Davis, of course, was having considerable success with his human relations text. Keith (then the consulting editor for McGraw-Hill, as I have now been for almost 20 years) agreed that a more behavioral science-based organizational behavior text was needed.

At the time, there was research and writing on organizational behavior starting to emerge in the scholarly literature. There were also research-based specialized books, such as Vroom's 1964 *Work and Motivation* and the just published (1968) Porter and Lawler classic, cognitively-based *Managerial Attitudes and Performance*. Also, before my book actually came out in 1973, I became aware of a couple of texts with organizational behavior in the title. For example, in 1968, Hampton, Summer, and Webber had come out with *Organizational Behavior and the Practice of Management* (Scott, Foresman), which was mostly readings and cases rather than a mainline text. Two other texts came out the following year. Joe Kelly came out with a OB text on the Irwin/Dorsey psychology list, and Al Filley and Bob House had a text published by Scott, Foresman, titled *Managerial Process and Organizational Behavior*, that was mainly a management book but had a few behaviorally-oriented chapters. Neither of these texts lasted beyond a couple of editions. I consider my text one of, if not the, first mainline organizational behavior text. At least, now in its eighth edition, it is the oldest OB text. Other first generation texts that are still alive and well, such as Hellriegel and Slocum (1976), came out a few years after mine.

The OB text took me almost five years, Sunday-Thursday evenings, to write. To this day, I still do the Sunday-Thursday evening routine for my writing. Unlike more recent authors, I had no other books to go from as far as conceptual framework or topic coverage. I drew from the combined psychology and management education I had received at Iowa, teaching psychology and the lessons I had learned, especially from Chris Argyris, while at West Point, and my new found behavioral orientation from my O.B. Mod. work to put this book together. Basically, the conceptual framework amounted to first laying down a historical, behavioral science (I had separate summary sections on anthropology, sociology, psychology, and social psychology) and scientific methodology foundation. Next, following the Argyris model of OB, consisting of the formal organization interacting with human beings, I had major parts on what I called "The Formal Organization System" and "Understanding Human Behavior." The organization part had three chapters on organization theory (classical, neoclassical, and modern), two on decision making, and one each on communication, control, and technology. The human behavior part had chapters on basic behavioral analysis (my O.B. Mod. orientation), perception, learning, motivation, and personality. The last part dealt with the outcome of the formal organization—human behavior interaction. There were chapters on group dynamics and informal organization, conflict and change, motivation and leadership for management, and behavioral applications to management.

I have put in considerable effort to revise the book over the years (every four years and, in more recent editions, every three years). This plain hard work of keeping up with the exploding research base, plus loyal adopters over the years, and having an excellent publisher in McGraw-Hill and their talented editors (Rit Dojny, who signed the book, Bill Kane, who brought me on as consulting editor, Kathi Benson, John Carleo, Kathy Loy, Alan Sachs, Dan Alpert, Lynn Richardson, and Adam Knepper, plus all the excellent support people) are the reasons for the sustainability of the book.

The upcoming edition (8th, 1998) retains the same basic conceptual framework but in a different order and, of course, different content. Besides a much briefer historical and behavioral science foundation (four chapters are now in one), the introduction now includes two chapters on emerging organizations (information-based, total quality and organization learning) and diversity and ethics. The parts now also flow from micro to macro instead of the reverse.

The twenty-five year change in content gives relatively less weight to macro units of analysis, such as organization theory and management functions, and relatively more weight to attitudes (e.g., organizational commitment and positive/negative affectivity), social cognitive variables (e.g., attributions and self-efficacy), and applications in motivation/job design, leadership and, especially in my case, behavioral management. Major OB topics that did not appear twenty-five years ago now include social learning theory, goal setting, teams, negotiation skills, stress, organization culture and international OB. Obviously, the motivation and leadership topics also include many new theories and research findings too numerous to mention. In other words, the OB field has certainly changed, but in this brief retrospective analysis, it really hasn't changed as much as we sometimes think.

These first 15 years at Nebraska were busy but fun times. I had a great boss in Henry Albers, a great colleague in Richard Hodgetts, great doctoral students such as Don White, Bob Kreitner, Tim Davis, Mark Martinko and Ken Thompson and, most of all, a great partner in Kay, and great children in Kristin, Brett, Kyle and Paige. We loved living and rearing our family in Lincoln. We have also had great social relationships, such as with long time accounting professor Tom Balke and his spouse, Karen. As couples, we still go out to eat and to a movie or sporting event almost every Friday and Saturday night.

Throughout this period, I had many opportunities to move on to bigger and supposedly better places. I took on the assistant dean's job at Nebraska right before turning 30 years old (they called me the "boy dean"), so I got that temptation out of the way early. I could only take two years of deaning and knew that I never wanted to do that again. However, because I had the administrative experience box checked, I have had many unsolicited calls and letters about dean's openings throughout my career. The same is true for head of department and distinguished chair positions. The only interviews I took were for the head of the management department at the University of Florida and a lucrative endowed chair at the University of Arkansas. I was offered both jobs at considerably more pay, and I liked the situation I would have had, but in the final analysis, Kay and I decided it would be best for our family to stay in Lincoln. Family has always come first for both of us. As it turned out, staying at Nebraska was the second best decision I have made (marrying Kay, as I said, was still the best).

RECENT YEARS AT NEBRASKA

In the mid-1970s both Albers and Hodgetts left Nebraska. In addition, Charlie Miller, the dean who had hired me, retired and was replaced by Ron Smith and his young associate dean, Gary Schwendiman from General Motors Institute. Ron soon moved on, but Gary took over and, up to last year, was my very supportive dean. He has now been replaced by my long-time colleague and close friend Jack Goebel. Although Jack was in accounting, he actually was interim chair of the management department for one year and then moved on to become vice chancellor and acting chancellor of the university before coming back as dean in 1995. However, the arrival of Sang M. Lee in 1976 as the permanent management department head really started a new era for me at Nebraska.

Sang, originally from Korea, had been at Virginia Polytechnic Institute for a number of years and was an internationally recognized scholar in management science. His major contribution was the development of goal programming. I had and still have tremendous admiration and respect for Sang on all dimensions—leadership, scholarship, humanity, and his racquetball and golf games. We immediately became and remain very close friends. I was extremely proud to serve as his best man when he married Joyce in March, 1991.

Sang provided me with tremendous support for my research and writing. Because to this day I still write everything in long-hand, I have greatly appreciated the word processing help I have received from the office staff, such as Linda Rohn, Debbie Burns, Phyllis Jacobsen, and especially Cathy Watson. Sang made the biggest contribution to my recent development by literally introducing me to the outside world. Sang Lee personifies the true meaning of a global scholar. Although Kay and I had traveled to Europe and Mexico on consulting assignments, our idea of international travel was going to Hawaii, where I taught summer school with the whole family along in 1976. Believe it or not, we did get tired of going to the beach everyday, and we were all happy to get back to Nebraska.

Starting over 15 years ago, Sang took me along for my first trip to Asia (Japan and to his family home in Seoul). We have been going back to Pacific Rim countries at least once a year ever since. In 1983, he started the Pan Pacific Business Conference. Every year Sang organizes this conference with a host Pacific Rim country through our management department office. With Sang as chair, Les Digman, John Slocum, Mary Ann Von Glinow, Clay Whybark, Bob Doktor, and I serve on the executive committee, and we have had a fantastic time at these conferences through the years. We have held very successful conferences in Japan, Korea, Singapore, Taiwan, Hawaii, Thailand, Australia, Malaysia, Canada, China, and New Zealand. In addition to these trips, I was in a number of cities in China just when the reforms were starting (many years before our conference in Beijing), have spent considerable time in Hong Kong, and have even taught a course at the university in Macau.

Not only did Sang add a significant international dimension to my career, but he helped me build up the OB area by hiring a young group of energetic and talented professors. Rick Mowday, Larry Pate, George Neely, John Cullen, Tom Mayes, Dan Ganster, Lew Taylor, and Bart Victor all started their careers at Nebraska. With my help Sang also brought in Les Digman to head up our strategy area. Les was a few years behind me at the University of Iowa, and I had worked for him as a trainer and consultant at the Army Management Engineering Training Agency in Rock Island, Illinois. Les has been a very valued colleague over the years, and he and his wife, Ellen, have become close personal friends with Kay and me. The same is true for Cary "Bud" Thorp, who was brought in by Albers after Bud had finished up at

Missouri under Bill Glueck. Sang had Thorp and Bill Torrence, before he retired, head up our human resources/labor relations area. Bud has not only been a competitive racquetball opponent for Sang and me but also a true friend and stable contributor to our department through the years.

The first crop of young OB colleagues eventually moved on to successful careers at other schools. All left for jobs they couldn't refuse. They were very productive while at Nebraska and, to this day, I believe are very positive about their Nebraska experience. Over the years they have been replaced by an equally energetic and talented group of young OB scholars—John Schaubroeck, Steve Sommer, and Doug May. I could not ask for a better group of colleagues. Until the arrival of Steve Sommer, I had, except for Hodgetts, almost always worked jointly with my doctoral students, rather than with colleagues. Steve and I work very well together (translated, he has great ideas and works very hard). Steve, Dianne Welsh, a former doctoral student, and I recently published an article concerning a study of Russian factory workers in the February 1993 *Academy of Management Journal*, and I consider it one of the best studies with which I have ever been associated. Steve and I continue to work closely together and should have some interesting articles out in the near future. Steve and his wife, Karen, are the proud parents of Erica who is about the same age as our first grandchild, Kourtney, so we have a lot of notes to compare on babies, as well. All of my colleagues in the department are very productive and keep me on my toes.

I probably worked more closely with doctoral students in the early days because I was the only OB professor. In more recent years I worked most directly with Avis Johnson, Diane Lockwood, Harriette McCaul, Nancy Morey, Jim Nimnicht, Stu Rosenkrantz, Carol Steinhaus, Robert Waldersee, Dianne Welsh, and Steve Williams and, more indirectly, a number of others with whom I have coauthored articles, including Ed Knod, Brad Chapman, Robert Koester, Jerry Sellentin, Chuck Millard, Todd Stewart, Barron Harvey, Nancy Carter, Janet Larsen, Barbara Kemmerer, LaVonne Wahl, Don Baack, Elaine Davis, Nancy Dodd, Marilyn Fox, and Linda Thomas. The most different of this group was Nancy Morey who came to me with a Ph.D. in anthropology. She obviously had a very different view of the OB world than I did, and I know I learned more from her than she did from me. Proof of this was the fact that, of the several articles we published together (including two in *Academy of Management Review*), they were all anthropologically-based.

I also collaborated on a number of interesting projects with Dianne Welsh. For her dissertation, she had taken the O.B. Mod. approach a step further with an intensive field study at a Burger King restaurant. She found that the Premack Principle (the use of high probability behavior to reinforce low probability behavior) could be successfully used to improve the performance of fast food workers. After receiving her Ph.D., Dianne went to Russia right before the break-up of the Soviet Union. She gathered some very unique and important information, but found it very difficult to get through the Soviet bureaucracy to obtain the data and take it out of the country. In a pioneering spirit, Dianne spent 16 hours a day for 12 weeks in the largest textile mill in the former Soviet Union gathering data. This Russian study replicated our previous O.B. Mod. interventions and a job redesign/enrichment study. Through a reversal experimental design, we found that the O.B. Mod. intervention (using both extrinsic rewards and just feedback and social/ recognition rewards) significantly improved the hard performance measures of the Russian textile workers. On the other hand, the culturally more complex job redesign intervention did not yield positive results.

To my knowledge, this Russian study was the first-ever experimental analysis of the impact of Western-based organizational behavior oriented interventions on worker performance in Central and Eastern Europe. Also, this study was important, to me, personally, because it covered the last base in my stream of research on O.B. Mod. Previously, we had demonstrated, at least to my satisfaction, that an O.B. Mod. approach could increase performance of employees in manufacturing, service, and health-care organizations. Now, we also had evidence that the approach held across cultures. Dianne also gathered direct observational data on the activities of the Russian managers at this huge factory to replicate the "Real Managers" study we had done a few years before.

This Real Managers study had come about because of a number of factors. First and foremost were my lingering doubts about the nature of managerial work as reported in our textbooks and the professional literature. In particular, my own extensive training and consulting work was with managers at all levels, functions, and types of organizations. I have worked directly or indirectly (through workshops/seminars open to the public) with supervisors and managers in almost all of the region's small, medium and large organizations in the manufacturing, service, and public sectors since I arrived at Nebraska. I have also had long relationships with very large nationwide organizations, such as Wal-Mart corporate training (for several years in the 1980s), and the National Rural Electric Cooperative Association (still going strong after 25 years with three groups of rural electric managers from across the country annually undergoing 6 weeks of training). In addition, I have my share of experience with firms outside the Midwest and internationally. Although I have worked and interacted with business heros (Sam Walton, for openers) and well-known companies (Nebraska has several world class companies, such as Union Pacific Railroad, ConAgra, Mutual of Omaha and Gallup), most have been first- and middle-level managers from all functions employed by small, medium and large manufacturers, banks, retail stores, fast food chains, state and local government agencies, and hospitals. That is, I have considerable experience with what I call "Real Managers," as opposed to the Lee Iacoccas or Jack

Welchs usually talked about in the professional literature or the sterile plan, organize, staff, direct, and control managers of the textbooks.

I had a problem justifying in my own mind (What mind? We should only deal with observable behaviors, shouldn't we?) that the managers I was dealing with were the same ones talked about in our literature. Mintzberg had cracked the door open with his *N* of 5 chief executive officers from large Boston and New York firms study, but this had just whet my appetite, not provided me with an answer. Also, I was concerned not only with what managers really do, but what do successful managers do (the fast trackers, those who rise quickly to the top of their organizations). Thirdly, I was concerned whether these successful managers were doing the same activities as effective managers (those who have satisfied and committed subordinates and perform well by getting quantity and quality in their units). These three research questions were very important to my own personal agenda, and I thought should also be important to the entire field of management. At least to my satisfaction, no theorist, research study, or reflective practicing manager had really answered these basic questions.

In my opinion, we have allowed the very insightful, but limited perspectives of the management pioneers, such as Fayol, Taylor and Barnard, the best-selling business heros, such as Lee Iacocca, gurus, such as Peter Drucker or Tom Peters, and even the early textbook writers, such as Koontz and O'Donnell or George Terry, tell us what managers do. I wanted a little more scientific evidence than their personal perspectives. The challenge was how to gather such evidence.

The easy answer to my research questions concerning the nature of managerial work would have been to do another questionnaire survey. I was and still am very concerned that our whole body of knowledge has been almost exclusively based on questionnaire gathered data. For example, the leadership field has greatly depended upon standardized questionnaires, such as the Leadership Behavior Description Questionnaire (LBDQ), whose validity has been criticized through the years. At Jerry Hunt's prestigious 1978 Leadership Symposia at Southern Illinois (What a wild ride it was out of that place. Chris Argyris and I "white knuckled" it together on the very bumpy ride to Chicago), I made two proposals in the paper I delivered: (1) We need to get back to observeable behavior as the unit of analysis, and (2) We need alternatives to questionnaire methods of measurement. In other words, I really questioned the value of another questionnaire survey.

My anthropology doctoral student, Nancy Morey, was getting my attention with her insistence on an emic approach (Let the setting and the subjects define the research situation, not the researcher) and the validity of direct observation data gathering and rigorous qualitative methodologies. I agreed with her, and Mintzberg had shown it could be done, at least with a very small sample of presidential level executives of large corporations. However, my generalizability concerns and my refusal to completely forgo statistical analysis dictated that I would need a fairly large sample to do this study correctly. The problem, of course, was that such a study, unlike the typical questionnaire study, would take a lot of time, money, talent and research site cooperation.

Once again, my situation at Nebraska allowed these barriers to be overcome. My O.B. Mod. research and writing had about run its course, and my textbooks were all established and required only revisions. I had the time, and it is a good thing because this study took about four years to complete. As far as the money went, Sang Lee had been subtlely urging me to get some grant money, and after doing some consulting work for Burt King at the Office of Naval Research (ONR), I submitted a proposal of my own. I received a grant of $225,000 to do the Real Managers study.

Once again, I found the talent among my doctoral students. In particular, Diane Lockwood had come out of the same master's program in interactive communication as Tim Davis had before her, and she did a great job in taking the initiative to develop the procedures and methodologies for the initial phases of the study. Diane and I trained the observers, validated the measures, analyzed the data and wrote the key paper describing the derivation of the managerial activities and the methodology used. This paper was presented at Hunt's 1983 Leadership Symposium held this time at Oxford University in England. It was at this conference that I got to really know John Slocum, and we have been close friends ever since.

There were no earth shaking results from the initial phase of the study. We found through free observation in the natural setting, and then using a Delphi process to categorize the data, that these Real Managers ($N = 44$) did the basic activities (communication, traditional management functions, and human resource management) suggested by our literature through the years. In addition, however, we did uncover a strong networking activity observed in these real managers. Networking in this study was behaviorally defined as interacting with outsiders and socializing/politicking. The second phase of the study used the derived behavioral categories made into a checklist that was filled out by trained participant observers every hour for two weeks on 248 real managers. This is where organizational cooperation was really needed, and we obtained it through personal contacts in organizations throughout Nebraska. In terms of relative frequencies of occurance, the activities were found to be as follows: traditional management (32%), routine communication (29%); human resources management (20%); and networking (19%).

Although I strongly felt that determining what managers really do was missing from our literature, the next phases on successful and effective managers were more exciting. These parts of the study were highly dependent upon the operational definitions of successful managers (an index of level over tenure was used) and effective managers (an index that combined subordinate satisfaction and commitment and perceived quantity and quality of the unit was used). Diane Lockwood had graduated and moved on to Seattle University after the initial phases, and Stu Rosenkrantz, who retired from the U.S. Airforce and was now working on his doctorate under me, took over most of the analysis. Avis Johnson, another doctoral student of mine, also assisted on the project.

We found, through both descriptive and inferential statistical analysis, that successful real managers did significantly more networking than their unsuccessful counterparts. The relative contributions of the various activities to managerial success were determined to be: networking (48%); routine communication (28%); traditional management (13%); and human resources management or HRM (11%). By contrast, the relative contributions the activities made to effective managers were as follows: routine communication (44%); HRM (26%); traditional management (19%); and networking (11%). This almost mirror opposite difference between the profile of activities of the successful and effective real managers was a very interesting and revealing finding. In fact, it was so intriguing that the Associated Press news wire picked up these results, and they were published in some newspaper stories across the country.

Next to my work on O.B. Mod. and my OB text, I consider this Real Managers study to be the most important work I have done. However, I feel that this huge study never really received the attention it deserved. We had several articles published from various facets of the study, and I had a summary piece in the *Academy of Management Executive*, which is still quoted quite a bit. I think it wasn't so much that the study was not accepted, as it was that the field either did not know about the study or did not really understand the findings. Much of the blame I place on publishing problems (a fundamental attribution error?). Richard Hodgetts, Stu Rosenkrantz and I incorporated the entire four-year study into what I consider to be a very readable book titled *Real Managers* (1988). Ballinger, our publisher, was in the process of building an impressive list but was suddenly swallowed up by HarperCollins, and our book, along with the others on the list, quickly went out of print. Academics, students and practitioners still frequently call or write to let me know they have just discovered this book (a collectors' item?), are very impressed, and would like to know more.

The Real Managers study carried me into the late 1980s. Toward the end of that project I was really getting heavily involved and interested in the international scene. International business texts were plentiful, there were a few international strategy casebooks, and Nancy Adler had come out with her paperback on international OB in 1986. Yet, going into the 1990s there were no international management books per se. Hodgetts, now at Florida International University for a number of years, called me to collaborate on such a book. I jumped at the chance.

I really was not too excited about the very specialized research that was going on in the OB field at the time. However, as I had been with my OB text, I was very excited about getting in on the ground floor of the international management field. As they had done over 20 years before with my OB text, McGraw-Hill saw the market need, took a chance, and signed what we believe to be the first mainline international management text. Parallel to when I wrote the OB book, Richard and I had nothing to go on, but we both enjoyed doing the book and learned a lot along the way. We have just completed the third edition, and there is a real challenge to keep up with this dramatically changing field. At this stage of my career, I needed something to get me charged up. The international management field, in general, and a project in Albania, in particular, gave me my second wind and I was off and running.

Through Sang Lee's effort to build the department through external funding, he landed about a million dollar grant from the U.S. Agency for International Development (AID) in 1991. At that time, AID was awarding competitive grants to universities to help the former communist countries make the transition to a market economy. Sang had traveled to, and had contacts in, the former Yugoslavia, so our department was awarded a grant to provide market economics education and management training to this leader of the Balkans. Almost simultaneous to our celebrating getting the grant was the disintegration of Yugoslavia and the outbreak of war among the various factions in the breakaway republics. At this point, the U.S. government shut down all aid programs with former Yugoslavia. We were held in limbo for several months. Then Sang got a call from AID to see if we could shift our project to Albania. This call had a dramatic effect on our department and me, personally.

After graciously accepting the change in plans, we rushed to the map to see where the heck Albania was and found it was a small country across the Adriatic Sea from Italy. Upon further investigation, we found that this relatively unknown country of about 3.3 million had a very unique, interesting and, in many ways, sad past. At the crossroads of history, Albanians are thought to be the only direct descendants of the ancient Illyrian people. Albania's rich cultural history has produced Roman emperors, their national hero Scanderbeg (who fought off the Turks) and, in modern times, Mother Theresa. Through the centuries, Albania has been mostly held hostage by tyranny, both foreign and, in recent times, domestic. Albania has been occupied by Byzantines, the Ottoman Turks, and, during World War II, by Fascist Italy and Nazi Germany. After the war, Albania came under a strict communist system ruled by the paranoid dictator Enver Hoxha.

For almost a half century, Hoxha totally isolated Albania from the outside world. He created an atheist state and, through his secret police, snuffed out any sign of political or religious dissent. During his reign of terror, it is estimated that several thousand were executed, and several hundred thousand were imprisoned, or forced into internal exile. By the time he died of natural causes in 1985, the country was devastated, both spiritually and economically. His legacy was to leave a beautiful country (mountains, streams, sand beaches, and plenty of natural resources) and beautiful people (proud, cultured, and educated) completely destitute. Albania is recognized as the poorest of the poor Eastern European countries. In the early 1990s, their per capita income has been about a dollar a day.

At the beginning of this decade, Albania's infrastructure totally collapsed. It was as if a nuclear bomb wiped out the country. One news account at the time noted that communism in most of Eastern Europe was dismantled carefully and painfully, but in Albania it was "ripped down, broken into bits, and carted away." Albania tried to recycle itself literally out of the rubble, but there was hardly a farm or factory that "escaped maiming as Europe's cruelest and most isolated dictatorship turned to dust."

After a few years of this total devastation, Hoxha's communist successor Ramiz Alia could no longer hold on. Albania became the last domino in Eastern Europe to fall when the people elected the democratic government of Sali Berisha in March of 1992. At this point, the U.S. government released the funds for our project in Albania. Before the Peace Corps office was set up or the others came (representatives from American and European agencies, a few courageous entrepreneurs, and evangelists), our Nebraska team made the first trip to Albania in May of 1992.

What an adventure and emotional experience this turned out to be. The Nebraska team (as we called ourselves and became known throughout the country) consisted of our leader Sang Lee, Les Digman (strategic planning), Robin Anderson (entrepreneurship), Harish Gupta (international economics), Ray Marquardt (marketing and agribusiness), Bob Echolt (a practicing business/accountant friend of ours who owns companies in Kansas City), and me. In 1994, Sang and I co-authored an article titled, "There Are Lessons to Be Learned as Albania Undergoes a Paradigm Shift," (*International Journal of Organizational Analysis*, Vol. 2, No. 1) and wrote, in detail, of our initial personal observations of Albania. I consider the article to be one of my best, and it is certainly the one which was written with the most heartfelt emotion. I really feel this account is worth reading, but let me just say that the whole situation was unbelievable. Yet, strangely to our friends and families, both Sang and I fell in love with the country and the Albanian people.

We were literally welcomed with open arms. We experienced firsthand the feelings our GI's must have had when they liberated oppressed people during the war. The Albanians, who had been taught to hate and fear Americans, not only were curious about this team of American business professors, they were extremely grateful and eager to learn everything. I was aggressively pursued to autograph my new edition of *Business* (co-authored with Hodgetts, published by Dryden) and began to feel like a "rock star" or Hollywood celebrity. They had no books of any kind (all the Albanian libraries had been stripped clean), let alone a four-color book that I told them had all the answers (just kiddingly) for the market economy. This text was provided to the hundreds of participants in our initial program and has just been translated into the Albanian language by the U.S. Information Agency. The treatment I received was in stark contrast with my undergraduate classes at Nebraska where the students typically ask, "What do we have to know from your book for the test?" The Albanians clutched the book to their chests, often with tears welling up in their eyes, and dutifully promised me they would read every word as soon as possible.

The first trip was the best, not only because we truly felt like pioneers representing our country and way of life, but because the small group of Albanians we worked with and got to know have become lifelong friends. In particular, both Sang and I, and the rest of the team and our families for that matter, have a close bond to this day with the Prime Minister's administrative assistant Timi Arbana and his spouse, Donika; the former Prime Minister's brother and now entrepreneur, Beni Bufi; my interpreter and prominent member of the opposition party, Georgi Kote; Reti Shutina, another interpreter now on leave from the World Bank to do her master's at the University of Toledo in the United States; Mimoza Laku, a school administrator from the seaport city of Durres; and truly outstanding young faculty members from the University of Tirana (all now in graduate programs at Nebraska) Beni Asllani, Silvia Trimi, and Laura Riolli. Laura was the only OB professor in Albania when I first met her in 1992; now she is one of my best doctoral students. These close friends were also part of a group that came to Nebraska in late fall of 1992 for 5 weeks of education and internships.

Through subsequent trips to Albania (12 for me so far and 14 for Sang) and the small groups of Albanians who periodically travel to Nebraska, we have made many more friends among the approximately 2,000 business enterprise and government agency managers and faculty members who have been participants in our program. For example, faculty members we have gotten to know very well from Tirana include Lindita Lati, Vladi Polena, and the former dean, Shyqyri Llaci; from the northern city of Shkodra, vice mayor Jozef Bushati and faculty members Alban Burazeri, Diana Sokoli, and Tila Sherifi; from the southern city of Vlora, faculty member Natasha Metaj; and from Korca, the dean of the business school, Frida Kosta, and the Dishnica family whose daughter, Iris, is now an undergraduate student (getting all "As") at Nebraska. Even though we have many new friends, it is the original core group members we see at social events or visit in their family homes on each of our trips, and we know we can call on them for anything we need, and they feel they can do the same to us.

Albania had bottomed out economically when we first arrived and then began to slowly recover. In fact, Albania is recognized to have had the highest growth rate in all of Europe in both 1993 and 1994. This growth, of course, has to be tempered when it is remembered that they were starting from below zero. However, Albania under Berisha's leadership did stabilize the currency, and their quick privatization of agriculture and housing can serve as the benchmark for transitionary economies. Post-communist Albania has also had a relatively stable government, although there are currently signs that this may be changing.

The real key as to why the Albanians seem to be on the right track in their transition to a market economy may be found in the attitudes and behaviors of the people themselves. The Russians and those in former Soviet Republics in Central Asia, that we also have been doing extensive work with the past couple of years, still seem to be living in the past to a large extent. According to a survey by the United States Information Agency, currently the Albanians are the only former communist country where the majority of the people feel they are better off now than before. Even though most Albanians are only earning a very meager income (when we first went to Albania our professor friends were making about $25 per month and now make about $50), the prices are the same as here in the United States, and basic health and public services are still terrible. Nevertheless, most Albanians still feel they are better off now than they were under the terrifying, oppressive conditions of the past.

Albania still has a long way to go and an uncertain future. We have the vantage point of being there from the beginning of the transition, and it has been most gratifying to see visible progress being made over the past four years. Unlike Russia and Albania's larger Balkan neighbors, we think Albania has a good chance to continue to progress if they can hold everything together. We have personally consulted with President Berisha about the problems. Specifically, the biggest problem we see for their future is to rebuild their infrastructure (health, education, the judiciary, transportation, roads, electricity, water, etc.) and get the value-added, job-producing manufacturing and service industries up and running. This, of course, requires massive infusion of capital, which has to date been largely absent.

A teaching assignment took me to the former East Germany in the summer of 1995, and I could see what progress has been accomplished in a few short years, as everything (roads, buildings, universities, public services, etc.) in East Germany has been or is being rebuilt with largely West German money. Thinking of the Albanians, I felt a little jealous and envious for them, because the Albanians were not as lucky as the East Germans and have had no one to help them rebuild. To date, the only visable signs of foreign investment in Albania are a Coca-Cola plant, a couple of very modern hotels in Tirana, and a small factory here and there. However, European and American agencies and the world financial institutions are trying to help, and I believe that projects, such as ours, where we are "teaching them how to fish, rather than giving them the fish" will be the most help in the long run. Still, the road ahead seems precarious and full of problems, and can be likened to the actual roads we travel on while in Albania: roads which are crumbling and full of huge potholes and are populated by speeding drivers in cars stolen from Italy, trying to dodge sheep and cattle, and the common people, who are merely trying to hold on to the little bit of the road that's left, as bad as it may be.

We currently are in the process of trying to extend our work in Albania, mainly by successfully starting an MBA program at the University of Tirana and more faculty exchanges. Whether we continue to be funded by the U.S. government or not, the Albanian experience has been very good for me, personally and professionally. I intend to continue to work for the Albanian people in some capacity. As Sang has been saying to me lately, it is time for us to give something back, and have fun doing it, and I wholeheartedly concur!

IT'S NOT OVER TILL IT'S OVER

The highlight of my professional career, and something I am unabashedly proud of, was when my peers elected me President of the Academy of Management in 1986. Besides being elected a Fellow of the Academy, the Decision Sciences Institute, and the Pan Pacific Business Association (only Sang Lee and John Slocum can also list all three), and receiving a number of distinguished teaching, leadership and service awards at Nebraska, I am most proud of being editor of *Organizational Dynamics* (*OD*) for the past several years. Working with the American Management Association, which publishes *OD*, and in particular Don Bohl, in recent years, has been enjoyable and professionally rewarding. I also continue to be excited about my consulting editor's job with McGraw-Hill and my new *Virtual OB* publishing venture with them. The college learning materials paradigm is changing. We think *Virtual OB*, which allows professors to tailor-make their textbook from our extensive database containing chapter modules, cases, readings and exercises, will be the wave of this future. Also, the database has the capability to go on-line, should this ultimate of electronic publishing become a reality.

Speaking of publishing in the future, I recently had a thrill that very few academics have been able to experience. Even though my two sons were not in the OB doctoral program under me, they both were in our related human resources management program, and I have recently had articles published that were co-authored with Brett (*Business Horizons*) and Kyle (*Human Resource Management Journal*). Although common in the medical and law professions, for some reason we academics don't seem to encourage our sons and daughters to follow us in our own or a closely related field. I think my sons would agree that, at least so far, it has worked out very well. In the future, I know I will greatly enjoy working with them both.

Not finally, I continue to enjoy working with my own current advanced doctoral students, such as Alex Stajkovic (he conducted a meta-analysis which found highly significant results of my O.B. Mod. studies over the past 20 years), Paul Marsnik (doing a learning/reinforcement-based analysis of innovation), Brooke Envick (did a replication of the Real Managers study with entrepreneurs), Steve Farner (doing an analysis of networking and internal customer service), and, indirectly, with Brenda Flannery and Rich Patrick. I also look forward to working with more recent doctoral students that I am directing, such as Alisa Mosley, Shanggeun Rhee, Laura Riolli, and Lena Rodriguez.

This autobiography has attempted to go back to the future with this "common man" in uncommon times. Such a disciplined look at the past as a prologue for the future has been very useful and rewarding to me. I recommend that everyone do the exercise. Two important themes emerged for me. First, I suspected, but now know for sure, that I was very, very fortunate to have wonderful people, too numerous to mention all by name, help me each step of the way. Second, I can now honestly say that I have no regrets. There isn't anything I would have, or should have, done differently. Being surrounded and supported by wonderful family, friends, and colleagues, in that order, and having no regrets, what more could anyone ask out of life?

PUBLICATIONS

1964

With M.S. Wortman. How many contracts ban discrimination in employment? *Personnel, 41*(1), 75-79.

With M.S. Wortman. Arbitration in a changing era. *Labor Law Journal, 16*(5), 309-315.

1965

With M.S. Wortman. The incidence of antidiscrimination clauses in union contracts. *Labor Law Journal, 16*(9), 523-532.

With M.S. Wortman. New evidence on contract clauses banning discrimination. *Personnel, 42*(5), 44-48.

1966

With M.S. Wortman. The prevalence of anti-discrimination policies in employment in unorganized firms. *Business Perspectives, 2*(2), 10-13.

1967

The faculty promotion process. Iowa City, IA: Bureau of Business and Economic Research.

Faculty promotions: An analysis of central administrative control. *Academy of Management Journal, 10*(4), 385-394.

Contributed to chapters on the art and science of military leadership. In S.H. Hays & W.N. Thomas (Eds.), *Taking command.* Harrisburg, PA: Stackpole.

1968

Management knowledge: An untapped resource for academic administration. *Advanced Management Journal, 33*(2), 83-88.

The impact of the Civil Rights Act on employment policies and programs. *Labor Law Journal, 19*(6), 323-328.

Training for qualifications: Solving the black worker's dilemma. *Training and Development Journal, 22*(10), 3-7.

A new look at antidiscrimination policies and programs. *Personnel Journal, 47*(12), 877-881.

1969

With M.S. Wortman (Eds.). *Emerging concepts in management.* New York: Macmillan.

Cases, readings and self-review for principles of management. New York: Wiley.

With R.M. Hodgetts. Managerial analysis of doctoral candidates and professors: Research attitudes and interpersonal relations. *Academy of Management Journal, 12*(2), 213-221.

With J.W. Walker & R.M. Hodgetts. Evidence on the validity of management education. *Academy of Management Journal, 12*(4), 451-457.

With R.M. Hodgetts. Government and business-partners in social action. *Labor Law Journal, 20*(12), 770-783.

1970

With J.W. Walker & R.M. Hodgetts. Who really are the promotables? *Personnel Journal, 49*(2), 123-127.

With R.M. Hodgetts. Women in management. *Personnel Administrator, 15*(6), 3-7.

1971

With D.D. White, Jr. Behavior modification: Application to manpower management. *Personnel Administration*, July-August, 41-47.

With R.M. Hodgetts. Managerial insights into campus unrest. *Business Inquiry, 8*, 5-12.

1972

With R.M. Hodgetts. *Social issues in business*. New York: Macmillan.

With R.M. Hodgetts (Eds.). *Readings in the current social issues in business*. New York: Macmillan.

Contemporary readings in organizational behavior. New York: McGraw-Hill.

With R.M. Hodgetts. Ecological challenge facing utilities. *Public Utilities Fortnightly, 89*(9), 23-27.

With W.E. Reif. Does job enrichment really pay off? *California Management Review, 15*(1), 30-37.

1973

Organizational behavior. New York: McGraw-Hill.

With R.M. Hodgetts. *Cases and study guide for organizational behavior*. New York: McGraw-Hill.

With R. Kreitner. The role of punishment in organizational behavior modification (O.B. Mod.). *Public Personnel Management*, May-June, 156-161.

The contingency theory of management. *Business Horizons*, June, 67-72.

With D. Lyman. Training supervisors to use organizational behavior modification. *Personnel*, September-October, 38-44.

With R. Ottemann. Motivation vs. learning approaches to organizational behavior. *Business Horizons*, December, 55-62.

1974

With W.E. Reif. Job enrichment: Long on theory, short on practice. *Organizational Dynamics*, Winter, 30-38.

With E. Knod. Job enrichment: Critical factors for successful implementation. *Atlanta Economic Review*, May/June, 6-11.

With T. Balke. The use of Delphi in setting objectives. *School Food-Service Journal*, June, 40-41.

With R. Kreitner. Behavioral contingency management. *Personnel*, June/July, 7-16.

1975

With R. Kreitner. *Organizational behavior modification*. Glenview, IL: Scott, Foresman.

With M.S. Wortman. *Emerging concepts in management* (2nd ed.). New York: Macmillan.

With B. Chapman. The female leadership dilemma. *Public Personnel Management*, May-June, 173-179.

With R. Ottemann. An experimental analysis of the effectiveness of an organizational behavior modification program in industry. In A.G. Bedeian, A.A. Armenakis, W.H. Holley, & H.S. Field (Eds.), *Academy of Management Proceedings* (pp. 140-142).

Management chapters. In K. Davis (Ed.), *The challenge of business* (pp. 43-77; 291-348). New York: McGraw-Hill.

1976

Introduction to management: A contingency approach. New York: McGraw-Hill.

With R.M. Hodgetts. *Social issues in business* (2nd ed.). New York: Macmillan.

With R. Koester. Experimental analysis of the impact of computer generated information on the choice activities of decision makers. *Academy of Management Journal, 19*(2), 328-332.

How PF/PR pays off for human resource managers. *Training*, December, 17-20.

With J. Sellentin. MBO in hospitals: A step toward accountability. *Personnel Administrator*, October, 42-45.

With M. Martinko. An organizational behavior modification analysis of absenteeism. *Human Resources Management*, Fall, 11-18.

With C. Millard & R. Ottemann. A new break through for performance appraisal. *Business Horizons*, August, 66-73.

With J. Sellentin. Management by objectives: Who is using it? Who isn't? *Health Services Manager*, June, 1-5.

An organizational behavior modification approach to organizational development. *Organization and Administrative Science*, April, 47-53.

How to apply MBO in the public sector. *Public Personnel Management*, March-April, 83-87.

With J. Sherman. Energy versus environmental protection: The dilemma facing electric utilities. *Journal of Behavioral Economics*, Spring, 29-41.

Motivation and productivity. In M. Gene Newport (Ed.), *Supervisory management* (pp. 173-190). St. Paul, MN: West.

1977

Organizational behavior (2nd ed.). New York: McGraw-Hill.

Contemporary readings in organizational behavior (2nd ed.). New York: McGraw-Hill.

With R.M. Hodgetts. *Cases and study guide for organizational behavior* (2nd ed.). New York: McGraw-Hill.

With T. Stewart. A general contingency theory of management. *Academy of Management Review, 2*(2), 181-195.

With R. Frame. Merging personnel and organization development. *Personnel*, January-February, 12-22.

1978

With M. Martinko. *The power of positive reinforcement*. New York: McGraw-Hill.

With T. Stewart. The reality or illusion of a general contingency theory of management: A response. *Academy of Management Review, 3*(3), 683-687.

With D. Lyman & D. Lockwood. An individual management development approach. *Human Resource Management*, Fall, 1-5.

With G. Neely. Using survey feedback to achieve enlightened AA/EEO. *Personnel*, May-June, 18-23.

1979

With M. Martinko. *The practice of supervision and management*. New York: McGraw-Hill.

With R. Kreitner. *Organizational behavior modification* (Spanish ed., A. C. Sanz, trans.). Mexico: Editorial Trillas.

With T. Davis. Leadership reexamined: A behavioral approach. *Academy of Management Review, 4*(2), 237-248.

With R. Koester. The impact of computer generated information on the choice activities of decision makers: A replication with actual users of MIS. *Academy of Management Journal, 22*(2), 416-422.

With T. Maris. Evaluating personnel programs through the reversal technique. *Personnel Journal*, October, 692-697.

With J. Schweizer. How behavior modification techniques can improve total organizational performance. *Management Review*, September, 43-50.

With T. Davis. Behavioral self management. *Organizational Dynamics*, Summer, 42-60.

With B. Harvey. Flexitime: An empirical analysis of its real meaning and impact. *MSU Business Topics*, Summer, 31-36.

Leadership: A proposal for a social learning theory base and observational and functional analysis techniques to measure leader behavior. In J.G. Hunt & L.L. Larson (Eds.), *Crosscurrents in leadership* (pp. 201-208). Carbondale, IL: S.I.U. Press.

Organizational behavior. In D. Yoder & H. Heneman, Jr. (Eds.). *ASPA handbook of personnel and industrial relations* (pp. 3-21 to 3-39). Washington, DC: Bureau of National Affairs.

1980

With R.M. Hodgetts & K. Thompson. *Social issues in business* (3rd ed.). New York: Macmillan.

Introduction to management: A contingency approach (Spanish ed., A.B. Montanez, trans.). Mexico: McGraw-Hill.

With T. Davis. A social learning approach to organizational behavior. *Academy of Management Review, 5*(2), 281-290.

With C. Millard & D. Lockwood. The impact of a four-day workweek on employees. *MSU Business Topics*, Spring, 31-37.

With D. Lyman & N. Carter. An accountability and appraisal system. *Management Review*, January, 46-51.

With D. Lockwood. Multiple ways to measure organizational change. *1980 Handbook for group facilitators*. La Jolla, CA: University Associates.

Resolved: Functional analysis is the best technique for diagnostic evaluation of organizational behavior. In B. Karmel (Ed.), *Point and counterpoint in organizational behavior*. Hinsdale, IL: Dryden.

1981

Organizational behavior (3rd ed.). New York: McGraw-Hill.

With K. Thompson (Eds.). *Contemporary readings in organizational behavior* (3rd ed.). New York: McGraw-Hill.

With R. Paul & D. Baker. An experimental analysis of the impact of contingent reinforcement on salespersons' performance behavior. *Journal of Applied Psychology, 66*(3), 314-323.

With K. Thompson & W. Terpening. The effect of MBO on performance and satisfaction in a public sector organization. *Journal of Management, 7*(1), 53-68.

With T. Davis. Beyond modeling: Managing social learning processes in human resource training and development. *Human Resource Management*, Summer, 19-27.

With M. Martinko. An experimental analysis of the effectiveness of a transactional analysis program in industry. *Transactional Analysis Journal*, July, 229-235.

Organizational behavior modification. In S.R. Michael (Ed.), *Techniques of organizational change* (pp. 47-88). New York: McGraw-Hill.

1982

With S. Lee & D. Olson. A management science approach to contingency models of organizational structure. *Academy of Management Journal, 25*(3), 553-566.

With T. Davis. An idiographic approach to organizational behavior research: The use of single case experimental designs and direct measures. *Academy of Management Review, 7*(3), 380-391.

With M. Martinko. Organizational behavior modification: Bridging the gap between research and application. *Journal of Organizational Behavior Management, 3*(3), 33-50.

With C. Snyder. Using O.B. Mod. to increase the productivity of hospital personnel. *Personnel Administrator*, August, 67-73.

Improving performance: A behavioral problem-solving approach. In L. Frederiksen (Ed.), *Handbook of organizational behavior management*. New York: Wiley.

1983

With D. Ganster & H. Hennessey. Social desirability response effects: Three alternative models. *Academy of Management Journal, 26*(2), 321-331.

With S. Rosenkrantz & H. Hennessey. Role conflict and ambiguity scales: An evaluation of psychometric properties and the role of social desirability response bias. *Educational and Psychological Measurement, 43*, 957-970.

With W. Maciag & S. Rosenkrantz. O.B. Mod.: Meeting the productivity challenge with human resources management. *Personnel*, March-April, 28-36.

With K. Thompson. Behavioral interpretation of power. In R. Allan & L. Porter (Eds.), *Power and influence*. Glenview, IL: Scott, Foresman.

Organizational behavior modification. In S.R. Michael (Ed.), *Techniques of organizational change* (pp. 47-90, Spanish ed., S.J. De Allub, trans.). Mexico: McGraw-Hill.

1984

With R.M. Hodgetts & K. Thompson. *Social issues in business* (4th ed.). New York: Macmillan.

With R. Kreitner. A social learning approach to behavioral management: Radical behaviorists mellowing out. *Organizational Dynamics*, Autumn, 47-65.

With A. Johnson & H. Hennessey. The role of locus of control in leader influence behavior. *Personnel Psychology, 37*, 61-75.

With T. Davis. Defining and researching leadership as a behavioral construct: An idiographic approach. *Journal of Applied Behavioral Science, 20*(3), 237-252.

With N. Morey. An emic perspective and ethnoscience method for organizational research. *Academy of Management Review, 9*(1), 27-36.

With D. Lockwood. Contingent time off: A nonfinancial incentive for improving productivity. *Management Review*, July, 48-52.

With J. Sherman. ESOP's: A win-win strategy for rewarding employees and employers. *Strategy and Executive Action*, Spring, 14-18.

With D. Lockwood. Toward an observation system for measuring leader behavior in natural settings. In J. Hunt, D. Hosking, C. Schriesheim, & R. Stewart (Eds.), *Leaders and managers: International perspectives of managerial behavior and leadership* (pp. 117-141). New York: Pergamon Press.

1985

With R. Kreitner. *Organizational behavior modification and beyond.* Glenview, IL: Scott, Foresman.

Organizational behavior (4th ed.). New York: McGraw-Hill.

With S. Rosenkrantz & H. Hennessey. What do successful managers really do? An observation study of managerial activities. *Journal of Applied Behavioral Science, 21*(3), 255-270.

With H. McCaul & N. Dodd. Organizational commitment: A comparison of American, Japanese, and Korean employees. *Academy of Management Journal, 28*(1), 213-219.

With N. Morey. Refining the displacement of culture and the use of scenes and themes in organizational studies. *Academy of Management Review, 10*(2), 219-229.

Organizational behavior modification and beyond: An interview. *Journal of the Norwegian Psychological Association*, Fall, 719-725.

1986

With J. Larsen. How managers really communicate: An empirical investigation of the relationship between communication behavior and managerial activities. *Human Relations, 39*(2), 161-178.

With R. Paul & L. Taylor. The impact of contingent reinforcement on retail salespersons' performance behaviors: A replicated field experiment. *Journal of Organizational Behavior Management, 7*(1/2), 25-35.

Structuring organizational rewards to motivate performance. In *National forum on human resources planning* (pp. 4-4 to 4-12). Washington, DC: U.S. Department of Defense.

With S. Rosenkrantz. Social learning theory and leadership. In A. Kieser, G. Reber, & R. Wunderer (Eds.), *Encyclopedia of leadership*. Stuttgard, Germany: Peoschel Verlag.

Fifty years later: What do we really know about managers and managing (Presidential Speech). *The Academy of Management Newsletter, 16*(4), 3, 9-10.

1987

With R.M. Hodgetts & K. Thompson. *Social issues in business* (5th ed.). New York: Macmillan.

With D. Baack & L. Taylor. Organizational commitment: Analysis of antecedents. *Human Relations, 40*(4), 219-236.

With C. Manz & K. Mossholder. An integrated perspective of self control in organizations. *Administration and Society, 19*(1), 3-24.

With B. Kemmerer, R. Paul, & L. Taylor. The impact of a job redesign intervention on salespersons' observed performance behaviors: A field experiment. *Group and Organization Studies, 12*(1), 55-72.

With K. Thompson. Theory D and O.B. Mod.: Synergistic or opposite approaches to performance improvements? *Journal of Organizational Behavior Management, 9*(1), 105-124.

With M. Martinko. Behavior approaches to organizations. In C.L. Cooper & I.T. Robertson (Eds.), *International review of industrial and organizational psychology* (pp. 35-60). New York: Wiley.

With N.C. Morey. Anthropology: The forgotten behavioral science in management history. In F. Hoy (Ed.), *Academy of Management Best Papers Proceedings* (pp. 128-132).

1988

With R.M. Hodgetts & S. Rosenkrantz. *Real managers.* Cambridge, MA: Ballinger.

Successful vs. effective real managers. *Academy of Management Executive, 2*(2), 127-132.

With T. Davis. Service organizational development. *Organizational Development Journal, 6*(4), 76-80.

With T. Davis. Organizational exit: Understanding and managing voluntary departures. *Personnel Review, 17*(4), 22-28.

With N. Morey. The use of ethnographic methodology for OD research. *Organizational Development Journal, 6*(3), 84-88.

The exploding service sector: Meeting the challenge through behavioral management. *Journal of Organizational Change Management, 1*(1), 18-28.

With D. Welsh & L. Taylor. A descriptive model of managerial effectiveness. *Group and Organization Studies, 13*(2), 148-162.

1989

With R.M. Hodgetts. *Business.* Fort Worth, TX: HBJ/Dryden.

Organizational behavior (5th ed.). New York: McGraw-Hill.

With R. Waldersee. What do we really know about EAPs? *Human Resource Management Journal*, Fall, 385-401.

With L. Thomas. The relationship between age and job satisfaction: Curvilinear results from an empirical study. *Personnel Review, 18*(1), 23-26.

With R.M. Hodgetts. Japanese HR management practices. *Personnel*, April, 42-47.

With M. Fox. Skill-based pay. *Personnel*, March, 26-31.

1990

With R.M. Hodgetts & K. Thompson. *Social issues in business* (6th ed.). New York: Macmillan.

With A. Johnson. The relationship between leadership and management: An empirical assessment. *Journal of Managerial Issues, 2*(1), 13-25.

With E. Davis. The healthcare cost crisis: Causes and containment. *Personnel*, February, 24-31.

With R. Waldersee. A theoretically based contingency model of feedback: Implications for managing service employees. *Journal of Organizational Change Management, 3*(1), 46-56.

Quality is an HR function. *Personnel*, May, 72.

With K. Thompson. Organizational culture: A behavioral perspective. In B. Schneider (Ed.), *Organizational climate and culture* (pp. 319-344). San Francisco: Jossey-Bass.

With T. Davis. Applying behavioral management techniques in service organizations. In D. Bowen, R. Chase, & T. Cummings (Eds.), *Service management effectiveness* (pp. 177-209). San Francisco: Jossey-Bass.

1991

With R.M. Hodgetts. *International management*. New York: McGraw-Hill.

With N. Morey. The use of dyadic alliances in informal organization: An ethnographic study. *Human Relations, 44*(6), 597-618.

With M. Fox & E. Davis. Improving the delivery of quality service: behavioral management techniques. *Leadership and Organization Development Journal, 12*(2), 3-6.

1992

Organizational behavior (6th ed.). New York: McGraw-Hill.

With R.M. Hodgetts. *Business* (2nd ed.). Fort Worth, TX: HBJ/Dryden.

With S. Williams. The impact of choice of rewards and feedback on task performance. *Journal of Organizational Behavior, 13*(7), 653-666.

With L. Wahl & C. Steinhaus. The importance of social support for employee commitment: A quantitative and qualitative analysis of bank tellers. *Organization Development Journal, 10*(4), 1-10.

With S.M. Lee & R.M. Hodgetts. Total quality management: Implications for Central and Eastern Europe. *Organizational Dynamics*, Spring, 42-55.

With R. Waldersee. A micro management approach to quality service: Steps for implementing behavioral management. In T.A. Swartz, D.E. Bowen, & S.W. Brown (Eds.), *Advances in services marketing and management: Research and practice* (pp. 277-296). Greenwich, CT: JAI.

1993

With D. Welsh & S. Sommer. Managing Russian factory workers: The impact of U.S.-based behavioral and participative techniques. *Academy of Management Journal, 13*(1), 58-79.

With D. Welsh & S. Rosenkrantz. What do Russian managers really do? An observational study with comparisons to U.S. managers. *Journal of International Business Studies, 24*(4), 741-761.

With D. Welsh & D. Bernstein. Application of the Premack Principle of reinforcement to the quality performance of service employees. *Journal of Organizational Behavior Management, 13*(1), 9-32.

With D. Welsh & S. Sommer. Organizational behavior modification goes to Russia: Replicating an experimental analysis across cultures and tasks. *Journal of Organizational Behavior Management, 13*(2), 15-35.

A paradigm shift in Eastern Europe: Some helpful management development techniques. *Journal of Management Development, 12*(8), 53-60.

With D. Baack & J. Rogers. Analysis of the organizational commitment of clergy members. *Journal of Managerial Issues, 5*(2), 232-253.

Meeting the new paradigm challenges through total quality management. *Management Quarterly*, Spring, 2-13.

With R. Hodgetts. U.S. multinationals' expatriate compensation strategies. *Compensation & Benefits Review*, January/February, 57-62.

With R. Hodgetts. U.S. multinationals' compensation strategies for local management: Cross cultural implications. *Compensation & Benefits Review*, March/April, 42-48.

1994

With R.M. Hodgetts. *International management* (2nd ed.). New York: McGraw-Hill.

With S.M. Lee. There are lessons to be learned as Albania undergoes a paradigm shift. *International Journal of Organizational Analysis, 2*(1), 5-17.

With R. Hodgetts & S.M. Lee. New paradigm organizations: From total quality to learning to world class. *Organizational Dynamics*, Winter, 5-19.

With R. Waldersee. The impact of positive and corrective feedback on customer service performance. *Journal of Organizational Behavior, 15*(1), 83-95.

1995

Organizational behavior (7th ed.). New York: McGraw-Hill.

With R.M. Hodgetts. *Business today*. Houston: Dame.

Virtual organizational behavior. New York: McGraw-Hill Primis.

With M. Rubach & P. Marsnik. Going beyond total quality: The characteristics, techniques, and measures of learning organizations. *International Journal of Organizational Analysis, 3*(1), 24-44.

With R. Patrick & B. C. Luthans. Doing business in Eastern Europe: Political, economic, and cultural diversity. *Business Horizons*, September-October, 9-16.

With B. Envick & R. Anderson. A proposed idiographic approach to the study of entrepreneurs. *Academy of Entrepreneurship Journal, 1*(1), 1-17.

With S. Sommer & S.-H. Bae. The structure-climate relationship in Korean organizations. *Asia Pacific Journal of Management, 12*(2), 23-36.

With S. Rosenkrantz. Social learning theory and leadership. In A. Kieser, G. Reber, & R. Wunderer (Eds.), *Encyclopedia of leadership* (2nd ed., pp. 1005-1021). Stuttgard, Germany: Schaffer-Peoschel Verlag.

1996

With S. Sommer & S.-H. Bae. Organizational commitment across cultures: The impact of antecedents on Korean employees. *Human Relations, 49*(7).

With P. Marsnik & K.W. Luthans. A contingency matrix approach to international human resource management. *Human Resource Management Journal* (in press).

With R.M. Hodgetts. North American management. In *International encyclopedia of business and management*. London: Routledge.

With R. Waldersee. Social learning analysis of behavioral management. In C.M. Johnson, B. Redmon, & T. Mawhinney (Eds.), *Organizational performance: Behavior analysis and management*. New York: Springer (in press).

With R. M. Hodgetts. Managing in America. In M. Warner & P. Joynt (Eds.), *International cross-cultural perspectives*. London: Routledge (in press).

With D. Bohl, J. Slocum, & R. Hodgetts. Peering into the crystal ball: Ideas and practices that will shape the future of management. *Organizational Dynamics,* Summer (in press).

Jeffrey Pfeffer

Taking the Road Less Traveled: Serendipity and the Influence of Others in a Career

JEFFREY PFEFFER

On August 15, 1989, in the ballroom of the Hilton Hotel in Washington, DC, I received the Richard D. Irwin Award for Scholarly Contributions to Management from the Academy of Management. In my brief acceptance remarks, I noted that performance was not just ability times motivation,[1] but performance in general, and whatever career success I had enjoyed in particular, was very much a function of one's specific social environment and social network—of the friends, peers, and colleagues who provided support, stimulation, ideas, and often provoked unanticipated changes in direction. During my life and career I have benefited greatly from the networks and organizations in which I have been embedded—but often the associations and their consequences were unanticipated and unplanned. This theme—of the importance of serendipitous social relationships and chance observations—is coupled with one other in thinking retrospectively about my career. I have often chosen a path not well-worn or traveled, and in so doing, have frequently adopted a perspective at variance with that dominating the field of study. I have seldom shied away from controversy. On reflection, of course, the two themes go together—for in order to traverse the road less traveled, one needs good friends, social support, and sources of insight that stimulate one to think in different ways.

Management Laureates, Volume 4, pages 201-233.

ISBN: 1-55938-730-0

AN ACCIDENTAL ACADEMIC

When I was growing up, I never planned to be an academic. My father was a small businessman (he owned a jewelry store) in Tucson, Arizona, where we had moved when I was six because of my mother's arthritis. Tucson then—and even now—had few professional opportunities or much commercial activity. It was the quintessential retirement town. I knew I would have to find a job elsewhere, and the only question was, "what?" Medicine had some intellectual interest, but I was squeamish and never keen on the sight of blood. Law, even then, seemed not stimulating enough, and those days, didn't pay all that well either. So, as a youth I recall developing a career plan to climb the corporate management ladder—preferably of a large and successful organization. After graduating from a private high school (the Webb School) in Claremont, California, I went to Carnegie Mellon University because it offered a program whereby one could get an undergraduate degree in Administration and Management Science and a master's degree in industrial administration in a total of five years, assuming one was admitted to the Graduate School of Industrial Administration. Also, Carnegie's engineering and science-oriented culture fit my academic proclivities—I had been strong in math and science in high school. So, off I went to Pittsburgh. My faith in my choice must have been unbounded, because I never even visited Pittsburgh or Carnegie before arriving to enroll. My choice was made somewhat easier by winning a Westinghouse Scholarship from Carnegie, which meant the financial drain of tuition would be less.

Things initially went according to my well-set plan. I earned high grades, was admitted to GSIA on schedule, and, because of advanced placement credit earned in high school, I received both my B.S. and M.S. degrees in just four years, graduating in the upper 10% of both undergraduate and master's classes. The only problem was the occurrence of two important events that began to alter my carefully planned career. One was the Viet Nam War. When I graduated in 1968, the interviewing process mostly consisted of asking prospective employers whether or not they thought they could get you a draft deferment. Other career and job dimensions seemed quite secondary in importance to those of us who didn't find serving in the armed forces an attractive prospect. This circumstance would affect my choice of my first job.

The second event was a talk on careers and career planning offered by Professor Alan Weinstein. Except for a paper with Srinivasan[2] examining the determinants of career success, I don't think I have ever read anything Weinstein wrote and I know he wasn't granted tenure at Carnegie. But his talk was compelling. He began by asking how many of us wanted to eventually become CEOs of major corporations. Of course, virtually every hand in the room went up. He then said, "Those of you who are women can put your hand down." There weren't many women even in the class then, but his perspective was that, in spite of the passage of the 1964 Civil Rights Act just four years previously, the amount of gender discrimination was too great to make achieving such a high-level position likely (ironically, one of the two women in my graduating Master's class later founded and was CEO of Quarterdeck Systems, a software company). Then he said, "If you are Jewish, put your hands down," going on to describe the religious discrimination that existed, particularly then, in numerous industries. Because I was (and am) Jewish, that got my attention. He then asked questions about how hard and for how long we were willing to work, how often we were willing to move—in short, a series of questions that assessed our willingness to sacrifice our lives for our corporate careers. I remember clearly that at the end of his series of questions, few hands remained up. Whereas the Viet Nam War affected my choice of a first job, Weinstein's talk would begin to shape how I thought about what I wanted to do and whether or not it was going to be feasible.

My career preferences were also being shaped by the sometimes random events that occurred during my job interviewing. For instance, in the process of interviewing for jobs, I visited with General Foods, in those days headquartered in White Plains, New York. I remember lunch clearly, for I was joined by a number of people from the Jell-O division. During lunch I learned more about that particular product than anyone could want, and also learned that in corporate America one often spent years mastering the minutiae of products like that—a finding subsequently given academic credence in John Kotter's[3] study of 15 successful general managers. Facing the prospect of terminal boredom in a corporate world interested in the number of extra jars in inventory of Maxwell House coffee and the marketing of Jell-O, the need for a draft deferment, and by now holding much more realistic career expectations, I took my first job with an Army think tank, the Research Analysis Corporation (RAC). My assignment involved working on a contract for the Assistant Secretary of Defense for Manpower to more accurately forecast draft requirements. I was given responsibility for overseeing the operation of a massive manpower simulation model and for doing what I could to improve the forecasting methodology. My innovation was to recognize that the number of voluntary enlistments in any given month were importantly affected by previous draft requirements. This not very startling insight, demonstrated empirically by analyzing existing data, substantially improved our ability to forecast the number of people that needed to be drafted. It also earned me a meeting with the legendary General Lewis B. Hershey, who had been head of the Selective Service virtually since its inception.

When I was called for my physical—my draft board in Tucson operating under the sensible principle of why make difficult decisions about deferments for people who might not be physically qualified anyway— I learned I would not be drafted because of my flat feet. I suddenly felt free to pursue any job or career option I wanted. While still at Carnegie I had interviewed with consulting firms and been turned down, so that door seemed closed. Now I had learned that working for a think tank full of Ph.D.s was going to be neither difficult nor exciting. It therefore seemed that a job in a corporation was not likely to keep me interested. During my year at RAC I had become quite friendly with Russ Fogler, a Ph.D. in finance from Columbia University working on cost studies for tanks, I think. Russ encouraged me to return for a Ph.D. and to go into academia. A number of professors at Carnegie had also encouraged this course of action while I was still there—among them Jim Laing and Ken Mackenzie, who taught organizational behavior, and Peter Frost, who taught finance. And, earning a doctorate could, perhaps, help me get a job at a consulting firm. So, influenced by my environment—friends and professors—I decided to explore returning to school.

SOJOURN AT STANFORD

Where should I apply? One place would be Carnegie, because I knew and respected it; I also applied to Stanford, because it was in California, and Berkeley, because it was also in the most desirable place to live in the world, the San Francisco Bay Area. I was admitted to all three, and have to admit I chose Stanford largely for its location. What should I study? I had taken more elective courses in finance than any other subject while earning my master's. But it seemed to me that, with the development of the theory of risk-return trade-off and the capital asset pricing model (for which my future colleague Bill Sharpe and his associates would subsequently earn the Nobel Prize in Economics), most of the truly challenging work in that field had been done. I had worked as a research assistant to Ken Mackenzie helping him run his communication network experiments[4] while at Carnegie and had also taken a number of organizations electives. This was a field with lots of remaining challenges, I thought, and implicitly and intuitively I recognized that this uncertainty presented more interesting career opportunities and better prospects.

So, off I went to Stanford. Jim Laing had graduated from Stanford's Sociology department, and he made sure I looked up Dick Scott when I got there. This accidental connection shaped my intellectual development profoundly. Mike Hannan had just joined the faculty in sociology, and both he and Scott came to serve on my doctoral dissertation committee. Meanwhile, the organizations program in the business school was somewhat unstructured. The school then (and to some extent now) operated under what came to be known as the "big boy" (subsequently, for obvious reasons, the "big person") theory. Doctoral students were presumed to be adults, the faculty were busy, and one was expected to determine one's own intellectual agenda. People were more than happy to meet with you and provide advice and guidance, but the student was responsible for taking the initiative, and if you drifted into oblivion, there would probably be little effort to locate you and return you to the fold. It is only a slight exaggeration to say that in some respects, I was largely self-taught.

I was interested in graduating as quickly as possible to get on with my newly discovered career—in fact I went through Stanford in two years—and knew I had to determine a particular focus for my work. The late 1960s was a time of ferment, excitement and change in organization studies. Thompson's *Organizations in Action,*[5] Katz and Kahn's[6] open systems perspective on organizations, and the work of the Aston group[7] were all changing thinking in the more macro part of the field, while the work of Weick[8] and Porter and Lawler[9] was influencing the more micro part. My associations in sociology tended to push me in a more sociological direction, and so too did some observations of the world—what Sutton[10] has come to call "closet qualitative research."

The late 1960s was a time of great ferment not only in the world of organizations but also in the United States. Not only was there the Viet Nam War and associated protests, but this period marked the beginning of serious enforcement of U.S. civil rights laws and the development of affirmative action programs for government contractors. How could one explain who discriminated and who didn't, and which organizations took affirmative action seriously and which resisted? It seemed to me that although the personality and personal values of senior management must certainly play a part, the critical factor was the organization's dependence (or lack thereof) on the government—that organization that was attempting to enforce those laws and regulations. Those organizations most dependent on the government would probably comply more, whereas those that were either independent or had countervailing power (because of the government's dependence on them) wouldn't. Thus was born a perspective on interorganizational power and dependence that has come to be called resource dependence theory.

If my dissertation research had proceeded smoothly, I probably would have become a great survey or field researcher, instead of developing an ability to find useful archival data. But it didn't. It turned out there were few organizations willing to give a graduate student access to their records or work forces to determine their dependence on the government and the race and gender composition of their employees. Fortunately for me, Yair Aharoni from Tel Aviv University was visiting at Stanford one year while I was there and we met and became good friends. He had collected data from a set of Israeli companies that included questions that could be used to assess their compliance with government policies favoring, for instance, investment in certain developing regions of the country, and the survey also had dependence data. So, I analyzed that and this became one chapter of my thesis and an early publication. One of my dissertation committee members was Gene Webb who, in the co-authored book, *Unobtrusive Measures*[11] had argued for a strategy of empirical convergence—taking the various empirical outcroppings of some theoretical point of view and examining them in multiple settings using multiple methods. The underlying logic was that while any single study would, inevitably, have some flaws, over time one could build a body of work that would prove to be consistent or inconsistent with basic predictions that emerged from the theoretical perspective. This became the strategy I followed in my dissertation.

It occurred to me that if I couldn't readily assess interorganizational influence, I might be able to examine organizational responses to interdependence. It was clear to me, again through some casual observation, that organizations, for good reasons, sought to retain their autonomy against attempts by other organizations to influence them. What might they do to retain discretion? One possibility was to engage in cooptation, for instance, by strategically using the composition of their boards of directors. Another possibility was to simply absorb the external dependence through merger. My dissertation, therefore, also consisted of a study of merger as predicted by patterns of dependence, the structure of boards from a random sample of corporations, and the size and composition of boards from a sample of electric utilities. And, the dissertation anticipated subsequent work on patterns of joint venture activity. Each of the separate empirical studies was quite consistent with the theoretical structure I was developing. Along with a chapter on unions as influencing organizations, empirically examining correlates of multi-employer bargaining—a study that was never published—my thesis developed resource dependence theory and examined it in a number of empirical manifestations.

HOW I GOT TO ILLINOIS

I entered the job market in 1971, coming from a prestigious school with what I thought to be an innovative dissertation. However, my thesis did not fit any established mold and neither did I. I like to tell people I was interviewed and rejected by all the best places—Harvard, Yale, Northwestern, the University of Chicago, and maybe some others I can no longer remember. I was told by some that, in studying mergers, for instance, I was doing economics—and not very well, at that. Paul Goodman, at that time at the University of Chicago, asked during our interview what my discipline was. "Organizational behavior," I replied. He repeated the question. After several cycles of this, I asked for the set of categories from which I could choose. That set consisted of traditional academic disciplines—Paul was trying to figure out whether I was a sociologist, a psychologist, a political scientist, and so forth. One of the strengths of my training in organizational behavior was its interdisciplinary focus. I was interested in studying organizations from multiple theoretical perspectives, and I told Paul this. Many of those interviewing me wanted to put me in a traditional, disciplinary box.

I wound up getting offers from Illinois, the University of Florida, and possibly one other place. At that time, Jagdish Sheth, a marketing professor, was the department chair at Illinois. The management department consisted of operations research, marketing, organizational behavior, and strategy. There were a lot of vacancies and Sheth was an ambitious, adventurous chair. Urbana-Champaign didn't look like California. But as I was deciding where to go, I learned that Gerald Salancik had accepted a position there. Salancik had worked with Gene Webb when both were at Northwestern's journalism school, and Gene had said wonderful things about him. With few alternatives that looked better, off I went to Urbana.

In retrospect, it was the most brilliant career decision I could have made, even though I clearly didn't really "make" the decision or even like it at the time. I recall being quite disappointed when Chicago, the last school to phone, told me they were not interested. The disappointment in my voice must have been palpable, and I wondered whether my strategy for doing such an ambitious and unconventional dissertation had been a mistake. Richard Hoffman (later at Rutgers) had made the call, and he provided some small consolation as well as some important advice. The consolation was that he thought Chicago had made a mistake. The advice was to decide whether I wanted to do what others wanted or expected, or to follow a set of research interests that I felt were important, understanding that this latter course might involve higher risk. So it was, very early in my career, I confronted the sometimes unpleasant consequences of doing things that were different. But I decided that one of the joys of being in academia was the opportunity to follow one's own judgment, and I resolved to continue to do so.

The good news was that my colleagues at Illinois were better than I could have hoped for anywhere else. Ken Rowland and David Cherrington were already there. Joining the faculty at the same time were Bobby Calder (now a full professor of marketing at Northwestern, who at the time had just graduated from North Carolina with a degree in social psychology) and Salancik, who received his Ph.D. from Yale. In marketing that year Illinois hired Peter Wright (now at Stanford) and Rick Winter. The following year we hired Barry Staw in organizational behavior. Probably 40 percent of the faculty in the College of Commerce had come in the previous three or four years, and with few other things to do in Urbana, there was a real sense of camaraderie and intellectual excitement. Most importantly for me, I found a set of colleagues who, in a very real sense, completed my training in both theory and research methods. And, I learned from the best—people like Staw and Salancik.

During my first year at Illinois, I began trying to publish the work from my thesis. Although I did have some early success with the empirical papers, my theoretical explication of resource dependence ideas was rejected by all the major journals in both sociology and organizations. I sent my dissertation off to Harper and Row publishers, and received a discouraging reply also. One day Salancik asked what I was doing with the manuscript, and I said, "nothing." He asked if he could read it. By that time I had pretty much given up trying to publish the theoretical ideas that formed the foundation for all of my empirical work. He approached me shortly thereafter and told me he found it interesting, and he offered to work on it with me. I accepted his offer. From that random event—he and I being at the same place at the same time, and my being so discouraged with attempts at developing and publishing this theory—came *The External Control of Organizations* (which included subsequent empirical work done after my dissertation) as well as years of collaboration and friendship with Salancik.

Although I was only at Illinois two years, it was a very important time in my development and in my life. I had an enormous amount of fun interacting with my colleagues. Because most of us had the same career history—too unconventional to be hired by the "right" places—we had an open, questioning, and innovative approach to the field. During my second year at Illinois, we hired Lou Pondy and Greg Oldham, although they started the following year after I had departed. My research during that period continued to develop along serendipitous lines.

Jag Sheth stepped down as department head after the first year and was replaced by Ken Uhl, also a marketing professor. Uhl thought the department should expand the number of students that we taught, because he believed this was a way to grow the faculty. It seemed clear to me and Jerry Salancik that if an increase in enrollment brought any increase in faculty at all, it was disproportionately small. We debated this with Ken, and finally decided to prove it to him. Illinois, being a public university, had publicly accessible budget and enrollment data. From the archives we gathered data on committee representation by department over time. And, we hired a research assistant to interview department heads about their perceptions of scarce and critical resources as well as their perceptions of departmental power. From our attempts to convince Uhl that power, not enrollments, drove resource allocations came our work on university budgeting and my continuing interest in power in organizations. At that time, a glance at an organizations textbook revealed that power, if it was mentioned at all, was discussed in terms of French and Raven's[12] bases of power. There was no consideration of power in decision making, where power came from, or how it was used. This line of research, started really to prove an argument, turned out to be both challenging and fruitful.

It was also at Illinois that I met and began collaborating with two wonderful graduate students, John Dowling and Huseyin Leblebici. John had come to Illinois to get his master's degree after working for the American Institute for Foreign Study, a travel and tour agency masquerading as a non-profit educational institution. His experiences with AIFS were instructive in understanding how organizations achieved legitimacy, and the work we did together on legitimation—a case study of AIFS—eventually appeared as both a journal publication and as part of the resource dependence book. John later went on to get his doctoral degree at Stanford. Huseyin, who first got his master's degree and later his doctorate at Illinois, helped Jerry and I with our work on decision making and uncertainty and subsequently collaborated extensively with Salancik on other projects.

While I was teaching at Illinois, I met Jan Beyer (at that time, Janice Lodahl) who was graduating from Cornell and entering the job market. Although we did not interview her, I became familiar with her work with Gerald Gordon[13] and immediately was captivated by the ideas. She and I became friends, and her work with Gordon on paradigm development, that I learned about through a chance contact during the recruiting process, came to stimulate a number of studies. One day while looking through the American Sociological Association's newsletter listing National Science Foundation grants, I noticed that the list was neither a random sample of sociology departments nor did it fit my sense of the prestige and productivity ordering of the field. That led Jerry Salancik and I to ask, as part of our interest in power, how uncertainty affected the decision making process. We concluded that in the presence of ambiguous criteria and information that did not permit one to clearly resolve a decision situation—such as what one would find in a low paradigm field—one would necessarily rely on social familiarity and identification—unambiguous cues that could help to resolve that uncertainty. We decided to study NSF grant allocations in fields that Lodahl and Gordon had found varied in their level of paradigm development. That study, when finally published, sparked both Congressional interest and also efforts by the Coles[14] and others to prove that less paradigmatically developed fields were not more particularistic in their decision processes.

ONWARD TO BERKELEY

During my second year at Illinois, Ray Miles from Berkeley called and invited me out for an interview. Although the job required me to take a cut in pay, I jumped at the chance to return to the San Francisco area. Back at Berkeley, my first research continued the trajectory that had been taking shape at Illinois. I was finally able to obtain joint venture data from the Federal Trade Commission, and with a research assistant, Phil Nowak, explored whether joint ventures also followed patterns of resource dependence. Once again, the issue turned out to be a timely one. There had been concern about joint ventures particularly in the oil and gas industry, and subsequently I had the opportunity to testify before a subcommittee chaired by Peter Rodino in the House of Representatives. The public policy implications of our work on joint ventures was published in *The Antitrust Bulletin*.

My interest in decision making, and particularly who benefited and who didn't from particular decision situations, led to a continuing interest in how various scientific fields operated. With two graduate students and in consultation with Janice Beyer—who was pursuing parallel but distinct work—I examined differences across scientific fields in the relationship between editorial board composition and publication in three journals, as well as the relationship between journal publication and prestige mobility, and how that varied by the level of a field's paradigm development.

I am sure my interest in power was encouraged by the times in which I was living (Viet Nam, Watergate) and also by my location on a very political campus, Berkeley. The study of resource allocations at Illinois had both methodological and substantive problems, and consequently I decided to replicate the study using two University of California campuses. A doctoral student, Bill Moore, and I gathered data on resource allocations to departments on both the Berkeley and Santa Barbara campuses over a ten year period. That data permitted us to replicate the Illinois findings about the sources and consequences of power while using better longitudinal research methods. The data also permitted us to explore the effects of power on the faculty salary determination process.

My research on organizational demography also began while I was at Berkeley, and the story of how that came about says a lot about learning from one's own experience, the importance of colleagues and friends in providing opportunities to develop one's ideas, and the role of chance in developing research directions. The observation was this: at the University of Illinois, assistant professors had a tremendous amount of influence. Each department in the College of Commerce had an elected advisory committee, and the college itself had an elected executive committee. I recall meeting at a colleague's house to plot slates of untenured candidates, and at the end of my first year at Illinois the assistant professors (who were quite numerous in all of the departments) had virtually taken over every elected position in the college. At Berkeley, there were few or no elected committees, and assistant professors were barely visible in the governance of the school or in almost any other way, for that matter. Because of my interest in university resource allocation and governance, not long after I arrived at Berkeley I had occasion to have lunch with Martin Trow, a distinguished full professor in the School of Education. When I reported my observed differences between the two places to him he immediately remarked, "But of course. Berkeley is a much more prestigious and high quality place, and so it is only natural that untenured faculty would have a much smaller role in such a place." Somehow the difference in quality didn't ring true as an explanation.

Berkeley at that time didn't have a fancy listing of the faculty, but each of us got a brief biography of our colleagues. Leafing through that one day I noticed that a significant number of the faculty had been at Berkeley quite a long time. And then it hit me—the difference between Illinois and Berkeley was in the age and tenure distribution of their faculties, and this must be part of the explanation for the difference in governance, culture, and other things. Of course, when I checked the literature to see if this factor of demography was used to explain how organizations worked, I found nothing, although there was some interesting and relevant work in sociology on cohorts. But this did not surprise me—by now I had learned that many of the variables and concepts I had come to find as being quite useful—power, resource dependence, uncertainty and paradigm development—were frequently not in the mainstream literature or even mentioned at all. And, also by now, I had developed experience that suggested that my observations and intuitions based upon them were often reliable in helping to uncover neglected explanatory factors. I had also developed a style which seemed to thrive on going after the unobvious.

But nothing came of my observation on demography for a while—I couldn't quite figure out what to do with it, how to test it, or how to develop it further, and I was busy with other things. Then in the late 1970s, as part of what I now see as a campaign to recruit me to Stanford, Jim March asked if I would participate in the work of the Committee on Aging of the National Research Council (NRC). The group was going to hold a conference and publish some papers, and they were interested in aging and organizations. I explained to him my undeveloped ideas about organizational demography and asked if this was close enough, and he said it was. So, I came to do my first work on organizational demography as a consequence of Jim's invitation to attend this NRC conference.

The work on demography also again demonstrated to me the costs and risks of doing things that were different as well as the operation of the psychological principle of selective recall. I proceeded to rewrite and expand the original demography paper into a theoretical exposition of the topic with some illustrative data, which I submitted to *Administrative Science Quarterly*. The paper was rejected. Having learned from what happened to the resource dependence work, rather than trying a lot of other journals, I expanded the paper and published it in the *Research in Organizational Behavior* series edited by Barry Staw and Larry Cummings. The paper came out in 1983 and, subsequently, the emphasis on demography has had significant impact on research and has proven to be a useful way of understanding organizations. One day, a while after the chapter had appeared, the telephone rang and it was Karl Weick, at the time the editor of *ASQ*. Karl was interested in understanding why good papers that were of interest to the journal had appeared elsewhere, and he was calling to inquire about the demography chapter. Why had I not published it with him? I told him why—he had rejected it. He couldn't believe it, so I wound up sending him a copy of the letter of rejection—signed by Karl E. Weick!

It was also during this time that Salancik and I completed *The External Control of Organizations*. Stanford had a post-doctoral fellowship program funded by the National Institute of Mental Health (NIMH) and at that time could give generous grants to visiting scholars. Jerry visited for a year while we completed the manuscript. At the same time, we also wrote our 1977 critique of need satisfaction models of job attitudes and a draft of our 1978 social information processing article. Once again, it was our tendency to lean against the prevailing theoretical views that inspired the work. The field was filled at that time with studies of the relationship between job characteristics and job attitudes. Salancik had a consistent theme about the organizations literature—there was very little organizational—e.g., social or contextual—and almost no behavior in the literature of organizational behavior. I suppose the intellectual bond that permitted someone who is often seen as more sociological and someone trained in Yale social psychology to work together was that we both believed in the importance of the social context and took the "social" aspect of organizations and social psychology seriously. In this, we somewhat anticipated the subsequent work on social networks.[15] Our first premise was always to try and avoid the fundamental attribution error and to look to the situation as at least an initial starting point for understanding behavior.

In the job design and job satisfaction literatures, that meant critiquing personality-based perspectives and looking to the informational environment—the social cues—as a way of understanding how people came to see their work and their attitudes. As usual, the work, departing from received wisdom and theory, was not necessarily well-received. Our reviews from *ASQ* were decidedly mixed, but Lou Pondy, then an associate editor, made the decision to publish both papers and to give our most critical reviewer, Clay Alderfer, a chance to write a critique.[16] Yet again, I learned that by leaning against the prevailing theoretical wisdom, when there were sound reasons for doing so, could result in initial difficulties in getting one's work out, but often led to that work having more influence when it was at last published.

THE MOVE TO STANFORD

I was happy with my colleagues at Berkeley, loved living in the East Bay, and had been granted tenure in 1975, two years after coming to Berkeley. John Freeman had joined the faculty, and the organizational behavior/industrial relations group was well-regarded in the school. By the late 1970s we had quite a bit of power and prestige in the school and, compared to many other organizations groups, had good security and status. But, I had come to Berkeley at a low salary, and the wage determination process, which operated under a set of quasi-civil service rules really made it difficult to improve one's position. And, the years of financial stress on the California university system were beginning to show in many ways. The social science library (shared by the business school) had almost no recent and significant books in it to check out. Its roof leaked and the ceiling had water stains. Barrows Hall, where the business school was located, was by repute the lowest cost per square foot building in the state, and it looked it. In 1977 I interviewed at Duke, received an offer, but decided not to go. John Freeman gave me great advice. Wait, he said. Duke had low-balled its offer, and rather than negotiate, I decided to see what happened.

During the 1970s, a group of us had begun to meet together regularly, funded first by a small grant from the American Sociological Association. The group consisted of Marshall Meyer, then teaching at Irvine, Mike Hannan, Dick Scott, and John Meyer from the sociology department at Stanford, John Freeman and I from Berkeley's business school, and Bill Ouchi from Stanford's business school. We were good friends and out of that collaboration came an edited book of chapters (Meyer et al., 1978). Bill Ouchi was coming up for tenure in 1979, and we all waited with anticipation to see what would happen.

My recollection was that I was giving a talk at Washington University in St. Louis the day Stanford voted to deny Ouchi tenure. I recall John Freeman calling me during the day in St. Louis, interrupting a meeting, to tell me. The speculation began as to what Stanford would do. There was a sense that they would need to replace Ouchi with someone of that cohort and age, and I was perceived as the logical choice. Stanford did call, and I received an offer and accepted it. Unfortunately, this chain of events for a while poisoned a good relationship with Bill. Even though I had obviously nothing to do with the Stanford decision—I was not yet a full professor and had not been asked to prepare an external letter of recommendation, and of course was not on the faculty to personally vote on the decision—it was still the case that I wound up taking his position, and even was assigned his office. It made things uncomfortable for a time, although eventually we have repaired the relationship.

The decision to move to Stanford was more difficult than in retrospect it probably should have been. I did have close social ties with a number of Berkeley faculty—Ray Miles, George Strauss, John Freeman, Joe Garbarino, Karlene Roberts—and breaking ties is difficult. But I had felt exploited by Berkeley. Berkeley had always claimed it couldn't pay a competitive salary. Confronted with an offer from Stanford and what I thought was an impossibly tight deadline, it matched that offer—which at the time represented a 69% raise from what I was then earning—in a week. That told me it could have always been more equitable, and having extracted the matching offer, I left. Jim March and Dick Scott were particularly influential in the recruiting process, as were my past dissertation committee members, Jerry Miller and Gene Webb.

There was another concern about the move, nicely captured by my Berkeley accounting colleague, Nils Hakansson. Nils had tried to convince me to stay, arguing that with the Stanford offer, I had credibility and power at Berkeley. To go to Stanford I would be a comparatively smaller fish because the pond was bigger. That was certainly true and moving to Stanford was, in many respects, a daunting experience. The Stanford MBA students enjoy a deserved reputation for being particularly demanding, and so teaching would be more challenging than it had been at Berkeley. And, as a new full professor, I would have no power or credibility. It would all have to be earned anew. But by this time I had grown accustomed to facing daunting challenges, beginning with my entry into academia and continuing with a set of research topics that tended to break new ground and which often challenged existing perspectives and provoked challenges. Salancik urged me to seize the opportunity, and I did. I joined the Stanford business school faculty in the fall of 1979. I had turned 33 years old that summer, and was at the time the youngest full professor in the business school.

AT HOME

I was almost immediately comfortable at Stanford. It was not just that I had graduated from the place. That had been seven years ago. It was that the style and ambiance suited me. Stanford at that time had a postdoctoral program in organization studies, which although not as financially generous as when Salancik had visited, was still quite attractive. A student of Barry Staw's, Jerry Ross, was interested in postponing becoming an assistant professor and taking on serious institutional responsibilities for as long as possible—a position I fully appreciated and understood. I was able to get him appointed at Stanford where he stayed for two years and he and I collaborated on a number of papers examining wage and status attainment processes. Like many of the people I have worked with, Jerry and I became good friends, and we still send each other birthday cards teasing each other about our advancing years.

Upon joining the Stanford faculty, I began to develop a new elective course in power. Although the topic was a good one, particularly for MBA students, my teaching skills were seriously underdeveloped. I had no real sense of how to decide on a good case rather than one that wouldn't work as well, was not particularly comfortable in a class discussion or case format, and certainly at that time had no appreciation of the power of video case material for teaching behavioral science. My teaching went all right, but I knew that I had to push the boundaries once more if I were to continue to grow and develop. I had mastered what I did and probably did it pretty well. Now I wanted to master some skills and a perspective that were totally foreign to me. So, within two years of joining Stanford, I went as a visitor to the Harvard Business School (HBS). I figured that was as different in teaching, research, and overall culture as any place I could imagine, and I was right. I was also right in terms of how that experience would help broaden my skills and perspective and, consequently, the course of my subsequent career.

OFF TO BOSTON

My year at Harvard was not always pleasant. I rented the house of a former Berkeley colleague then teaching at MIT, Rick Bagozzi, in Winchester. I looked at the house on a Sunday, and carefully clocked that it was only about 6.5 miles from HBS. Unfortunately, my California experience—where time equals miles—didn't translate too well in Boston. That year they were doing a lot of road construction—on the worst day it actually took one hour to get in, but on average it was frequently close to 40 minutes. I arrived before Rick and his wife were ready to leave for their year in Germany, so I spent the first two weeks at Harvard in Jack Gabarro's basement. Jack was also on leave and Bob Eccles was renting his house for the year. Bob was kind enough to let me stay in the basement, which wasn't as bad as it sounds, until I could move out to Winchester. Based on that and other experiences as outsiders—Eccles as an untenured faculty member and me as a visitor—we became good friends and are to this day.

John Kotter had also just recently gotten tenure at Harvard and was also developing an elective course called Power and Influence. My assignment was to teach a doctoral course in the fall. Both John and I would then teach two sections of Kotter's course in the spring. Teaching Kotter's course was just that—he gave me the syllabus, we discussed it, he took none of my suggestions, albeit with his typical grace and charm, and he gave me the final syllabus to teach. He also met with me regularly to talk about the upcoming cases, how the course was going, his philosophy of the course and teaching, and, in short, served as an extraordinary mentor and guide through the Harvard experience. The course met for 16 weeks twice a week—32 class sessions times two sections meant 64 classes. I learned to teach cases by doing it 64 times—with Harvard students and with Kotter's great advice. It was learning by doing. I acquired a much better sense of how to judge case material and I was exposed to video case material. Some of it I still use, but more importantly, I became a fan of that style of pedagogy—so much so that a nickname I have acquired at Stanford is Captain Video.

I used my opportunity to teach a doctoral course at Harvard to begin to launch yet another new line of research and teaching, this dealing with the connection between organizations and labor markets. I had written a paper while still at Berkeley trying to draw links between organizational processes and salary attainment. It seemed obvious to me that careers were embedded in organizations and that wages were determined by organizations. Therefore, studying careers or salaries without understanding the organizational context and processes that affected careers seemed to me to be unproductive. At the same time, I was convinced that income and careers were important outcomes—people worried about them a lot. Therefore, understanding organizations would be enriched by focusing on career and labor market processes in organizations because these were significant organizational phenomena.

Harvard and Boston were big adjustments. At Stanford, on the week-end one can see George Shultz, former president of Bechtel and secretary of labor and treasury, walking around in blue jeans. Even during the week, many of the Stanford faculty dress casually. I knew Harvard was more formal, but somehow, I was always one level below where I needed to be—if I wore a sweater and slacks, I should have worn a sport coat; if I was in a sport coat and tie, which is what I wore when I taught, others would be in suits. And if I wore a suit, I could be sure that a tuxedo was the dress needed.

Two incidents particularly stand out during my year at Harvard. In early April Boston had a terrible snow storm. I had worked at home that day. Although the Bagozzi house had much to recommend it, it had no garage, a very long driveway, and no snow blower, because the preceding winter (Rick's first in Boston) had been exceptionally mild. Moreover, when it snowed a lot with wind, the snow drifted in front of the front door so one couldn't get out except through the basement, which opened in the back of the house. That evening I called John Kotter to ask him if we were having class the next day—all of the rest of Boston was closed. He said, "Of course," but if I couldn't make it in not to worry. We taught in the same room, the same material, alternating class sessions with him going first. If I didn't show, he would just do my class. I pondered what to do. The wind was blowing ferociously and I could see that getting out the next day would be difficult—requiring me to dig a path down the driveway, break through the huge mound of ice and snow created by the street plowing, and then take a bus into Harvard. It looked daunting. But then I realized that if I didn't show, students (and probably professors) would talk about the weenie from California (and Stanford at that). So at 4 the next morning, I began my task of shoveling my way out. I would shovel for a while, get tired and cold, go back in the house, and then reemerge to do battle with the elements. I made it to class on time to the surprise, I think, of Kotter and to a round of applause from the students. They were all mostly there because most of them lived on campus. I told them that I did this once to show them I could and that Californians were tough, but would not do it again (should the need arise), to show them I wasn't that stupid.

The second, and more important, incident had occurred earlier in the year. Stanford was hiring that year and I, of course, wanted to be somewhat involved but was out of the loop. Fortunately, one of our leading candidates, Jim Baron, was being invited to talk to the Harvard sociology department. Harrison White invited me to lunch, and I got to know Jim. As White was paying the bill for all of us it dawned on him that he had probably just paid for a recruiting lunch for Stanford. Baron's interests in organizations and labor markets not only fit my own emerging interests, I was convinced this was an important way to begin doing more interesting work in the area of human resources. I was glad when Stanford offered Baron the job and was thrilled when he accepted. It turned out my location in the East helped in my recruiting. I would teach my first class of the day and get back to my office at about 11:30 and call Jim in California. In those days, with no children, Jim tended to sleep late. I would invariably wake him up. Finally he said, "what do I have to do to get you to stop calling so early." I said, "accept Stanford's offer." He did. I don't think I have awakened him since.

BUILDING AND BROADENING

I had been hired to help build the organizations group at the business school and, in particular, to build its strength in the more "macro" part of the field. At the time I joined the group consisted of Harold Leavitt, Gene Webb, Jim March, Joanne Martin, and Jerry Porras, the latter two untenured at the time. The first person we hired the first year I was on the faculty was Don Palmer, a sociologist out of SUNY-Stony Brook with a Marxist theoretical orientation and an interest in interlocking directorates. Don's work on the effect of control on organizational structure and headquarters location decisions and his personal teaching style earned him quick acceptance in the business school. He remains a good friend. Jim Baron was the next addition to the group. Early on he developed an elective on human resource management, and his hiring really marked the beginning of the school's subsequent expansion into that domain. Jim has been a collaborator on two articles and a very close friend. His work on the employment relation has influenced my thinking profoundly, and hiring him was one of the best decisions Stanford has made. Since those early additions Stanford hired Rod Kramer, Joel Podolny, Michael Morris, Charles O'Reilly, Pamela Haunschild, and Maggie Neale. Furthermore, the business school assisted in recruiting Mike Hannan back to the sociology department and offered him a joint appointment, and the organizations group helped in recruiting Bill Barnett into the strategy group. Except for Palmer, all of the untenured faculty who have come up for tenure (Martin, Porras, and Kramer) have earned it, and the group has developed an excellent reputation in the school for hiring and developing outstanding junior faculty. The group has flourished, not only expanding in size but also in the strength of its doctoral program and its administrative contributions to the school (for instance, Webb, Porras, and Baron have all served as associate deans).

Much of our success derives from a culture or set of norms that says that an important responsibility of senior faculty is providing advice and guidance to our junior colleagues. Another element of the culture that has developed is the sharing of advice and information about teaching, so that the transition to Stanford's comparatively demanding classroom environment becomes less taxing. Over the years I have come to appreciate the importance of these activities and have tried to be helpful. Although the group is quite diverse both in intellectual style and disciplinary background, we have understood the importance of being united in the face we present to the rest of the school, and have largely been able to maintain a unified stand with respect to our external relations.

This institutional leadership role has at times been problematic for me. Although I think I have been reasonably effective, one of the presumed joys of academia is the absence of task-related interdependence and the consequent ability to do one's own thing. Playing a leadership role invariably involves spending time on activities inside the school and the group. I think people don't always appreciate that holding and exercising power and influence involves serious work, and that those with influence find many demands placed upon them as a consequence of their status. Stanford's culture has been dominated by the economic model of human behavior and is, as a consequence, quite individualistic. Thus, on the one hand, my activities on behalf of the community take time from my own work, but on the other hand it is unclear that there is a very favorable trade-off between the potentially adverse consequences for individual productivity from time spent on behalf of the collective. These tensions have been, nevertheless, useful for me as they cause me to continually revisit questions of how I spend my time and what I believe to be enjoyable and useful activities to engage in.

My year at Harvard and the move to Stanford also caused me to rethink my relationship with the world of practice—which up to that point had been pretty minimal. I had gone to Harvard to see at close range an alternative model, not only for teaching and research but also for how a business school relates to its professional constituency. Although I did not fully agree with every aspect of Harvard's approach, it seemed clear to me that I could learn a great deal from the world of practice if I got more exposure to that world. Thus, the 1980s saw an increased interest and involvement in executive education and consulting. My enhanced teaching skills made me a fairly decent executive education instructor, and the Stanford platform made many more opportunities available.

What I mostly learned from this experience was simply this: that although most of our theories and research concern what might be called "content"—what to do about organizational design, decentralization, self-managed teams and quality circles, job design and the division of labor, and so forth, what managers really have problems with is implementation—the "how." This insight, which I am sure is obvious to virtually all managers and most of my colleagues with more practical experience than I, is nevertheless hard to keep in mind, because the pressures to study what affects what often drives out concern for issues of how to implement things.

The concern with implementation was, of course, related to the course on power and politics in organizations I had been teaching at Stanford. But it caused me to reframe the course, emphasizing clinical skills as well as conceptual knowledge. The reframing was successful, and the course initially grew to two and now three sections, one of the most popular in Stanford's MBA elective curriculum. I have been particularly pleased that other colleagues, Gene Webb and Rod Kramer, have been interested in the course and its ideas and have largely taken over teaching it. And, this concern with implementation also helped stimulate me to write a book that might be read by practicing managers, *Managing with Power*.

I had written *Power in Organizations* as a foundation for the elective course and to provide a summary treatment of the literature on power, including my own work. Unfortunately, it was not written from a reader's point of view, unless the reader was an academic. Over the years, student complaints grew, particularly after others began teaching the course—it is probably easier to complain about a book behind an author's back than to his face. My colleague and friend, Hal Leavitt, had been encouraging me to write something for a broader audience. I took this encouragement as a compliment reflecting the fact that he thought I had interesting and important things to say. Finally, the day came that Gene Webb told me he was not going to use the book that quarter when he taught the course. Gene had been on my dissertation committee and was a good friend, so I figured it must be really bad. At the end of the quarter, I asked Gene how things had gone. All right, he had replied, but the students weren't happy about not having a text for the course. So, I learned, they wanted a book covering the topics, but not the one currently available. One other event helped push me to decide to try my hand at a more popular but still sound treatment of the subject of power in organizations. In 1985 I had met and in 1986 had married Kathleen Fowler, a second marriage for both of us. Kathleen had never dated or known an academic and seemed somewhat fascinated with the academic life, which she referred to as a "black hole in the job market." She wanted to read something I had written. The first thing I gave her was the first book on power, thinking that was about as readable and interesting as anything. She read it all, I think, but also then began encouraging me to write something more accessible.

I found writing *Managing with Power* enjoyable. I had been an editor of my high school yearbook and had actually done quite a bit of non-scientific writing earlier in my life, and I found getting away from an academic formalism while trying to maintain rigor and integrity of ideas to be both challenging and fun. I was particularly pleased by the reception the book received—not so much by its sales but by the regard it garnered from colleagues, students, and practicing managers.

I determined to follow up that book with a second, *Competitive Advantage Through People*, that would capture my thinking, as developed in both an elective course I and others had taught, and my own research. I hoped that book would be of a similar style—to use Kathleen's apt phrase, "dumbed down just enough." My growing research and teaching interest in human resources once again illustrates the importance of accident and personal ties. I have already mentioned the hiring of Jim Baron, who later came to serve as head of the human resource initiative in the business school. Several other events helped crystallize my direction, as well as cement my collaboration with Baron.

In 1985, Tom Peters, a graduate of Stanford's MBA and doctoral programs, gave his friend and mentor Gene Webb some funds for a study trip to Sweden, where Lennart Arvedson was working with a Swedish social policy research institute. Lennart had been a doctoral student with me, actually sharing an office our first year in the program, and was later to work for Peters in Europe, so social connections loom large in the tale. To make the trip less of a total boondoggle, it was determined that everyone would write a paper, although collaboration was permitted, on something that had to do with labor markets or labor policy. Baron and I collaborated on what became the paper "Taking the Workers Back Out," a review of the trend toward externalizing the employment relationship. The working title of the paper was "Ticket to Sweden." That paper and its reception when subsequently published in the *Research in Organizational Behavior* series convinced both of us that there were important issues in the management of the employment relation that were being overlooked.

But, to some extent that work already followed an interest in organizations and labor markets. Following the doctoral course I had taught at Harvard, I returned to Stanford and began a series of collaborations with postdoctoral fellows, each of whom became good friends and who helped my research enormously. The first was Yinon Cohen, who Don Palmer had known from the doctoral program at Stony Brook. Yinon discovered some data that had been gathered on the employment practices of Bay Area establishments by the Institute of Industrial Relations at Berkeley. He was able to acquire data that proved useful in understanding internal labor markets and organizational hiring standards. Then came Deidre Boden, from University of California, Santa Barbara, who introduced me to ethnomethodology in general and conversation analysis in particular. She was followed by Alison Konrad and

Nancy Langton, and then the postdoctoral program, always funded by NIMH, ceased. Alison was interested in gender discrimination, and Nancy brought an interest in labor markets that we were able to explore using Carnegie Foundation data on faculty salaries. Cohen, Konrad, and Langton all collaborated on various research projects that dealt with organizations and labor markets, and their availability and interest stimulated that particular direction of my work, already begun with Jerry Ross.

At about this same time, Alison Davis-Blake was a student in the doctoral program also with an interest in labor markets. Our collaboration, begun while she was a student and continuing thereafter, has been extremely productive. We have examined the effects of the proportion of women administrators on salaries of both male and female college administrators, the causes and consequences of wage dispersion among college administrators, and wrote a paper critiquing dispositional approaches to organizations and the resurgence of interest in personality. This latter paper was a return to some of the themes pursued with Salancik more than a decade previously—an emphasis on the social causes of behavior and a substantive and methodological critique of trait-based perspectives.

LEARNING IN A CRISIS

I had earned my doctoral degree in a little more than two years, earned tenure at Berkeley while in my twenties, and come to Stanford as a full professor in my early thirties in part because of a sense of limited time. Heart disease ran on both sides of my family—male cousins from both sides had died in their early 40s, my mother had died from a heart attack at the age of 62, and since childhood I had elevated (but untreated) cholesterol. I had wanted to attain success early enough to be able to enjoy it, and in large measure, I had. In the summer of 1991, walking across campus to the faculty club, a distance of about a mile, I felt out of breath. I attributed it to poor conditioning—although I walked fairly regularly, I did not engage in strenuous aerobic exercise. In June, 1993, Kathleen and I visited Belgium and Maastricht just over the border in the Netherlands, where Hans Pennings was spending the summer and where I spent a day giving a seminar to some doctoral students. We had eaten a large (and probably too rich) meal, and were walking back to the hotel when a thunderstorm came up. As we walked faster to get back, I felt out of breath in a different way than I ever had before. I had just had the flu, and attributed it to that, but in the back of my mind I knew something was not right. By the end of the summer of 1993, the symptoms were quite severe, and I determined to tell my doctor about them on my next visit in September. Meanwhile, Kathleen and I went to Australia and New Zealand on a combination of vacation and work, where in the last week I was in five cities in six days.

Upon my return, I went in within two days for my already scheduled doctor's appointment. My doctor, who given my family history took all of this seriously, ordered a treadmill test for me. It was Wednesday, September 15, and we were scheduled to attend Rosh Hashanah services that night. Around 7 the doctor called and told me that "the professor had failed the test." He was concerned by the results and felt that I should have an angiogram very soon. He also said I should do nothing to stress myself, which I replied, certainly meant not talking to him. One week later, the angiogram revealed that I would need bypass surgery, for the blockages could not be treated by angioplasty. Fortunately, I have a cousin, David Lehr, who has been a very prominent cardiologist in Miami Beach. Although we had not been close, I tracked him down and he agreed to look at the films. His conclusion was the same as all the others—that if I had a heart attack, I would die instantly and given the extensiveness and location of the blockages in the coronary arteries, I should do something about this soon.

We had been scheduled to visit INSEAD in mid-October, but that trip was postponed. Mitch Koza was quite nice when I called to cancel, noting I had probably given him the most legitimate excuse he had ever received, and promising to see if the trip could be rescheduled in the spring if I felt up to it. During the next three weeks I learned a lot about the treatment of heart disease and the practice of medicine in California, where heart surgery still tends to use veins rather than arteries that have a much better chance of resisting subsequent deterioration. We chose a heart surgeon and a hospital (both in San Francisco), convinced the surgeon to do the operation the way doctors in Miami had advised, arranged for him to get me a private room, and in short, "managed my doctors with power," as one colleague noted. On October 11, 1993, I underwent bypass surgery and although certainly not pleasant, I was tremendously pleased with both the doctors and the hospital. Four weeks later, Kathleen and I went out to dinner to celebrate her birthday, and earlier that day I had driven down to Palo Alto—the first time I had driven at all. In March, we took our postponed trip to France and INSEAD, celebrating my complete recovery.

This experience reinforced my appreciation for the relevance of the material we teach—understanding organizations helped immensely in negotiating the world of medical care in the 1990s. And, it caused me to once again understand how important social relationships are—the concern of friends and colleagues (at one point the hospital switchboard stopped taking calls because there were too many) helped immeasurably in confronting a difficult situation. Finally, this experience gave me the opportunity to reflect, once again, on what I was doing and how I was spending my time. This reflection renewed my appreciation for the pleasure I get out of my job.

NEW COLLABORATIONS

My growing interest in implementation, or lack thereof, of high commitment work practices led me to accept an invitation to participate at a conference at the Industrial Relations Research Centre at Queen's University and to prepare a paper on the various barriers to implementing what we know about how to manage work. Some, although not all, of those barriers implicate social psychological processes, such as commitment and consistency, which tend to encourage managers to justify their efforts at control and management. My ability to pursue these social psychological issues suddenly was enhanced when Robert Cialdini decided to spend his sabbatical year at Stanford. Because of Gene Webb's respect for his books on influence, Bob had been a visitor to Stanford on occasion during the summer. With him here for the year, I used the opportunity to see if I could get him interested in issues of why people don't do what they know they should. I had some insight, but was quite convinced that I needed an imaginative experimentalist in order to implement the ideas. Anyone who has followed Cialdini's career knows he has been interested for some time in how to undermine the undermining effects of extrinsic motivation and surveillance and that he has an interest in real world issues. So, he was interested in the problems I posed, and we have just begun a collaboration that I hope will extend over several years and produce both theoretical understanding and some practical suggestions for helping organizations implement better ways of managing more consistently.

In building still more and new collaborative relationships, I find myself still trying to learn from others and to put myself in situations where I have the opportunity to do so. I find myself still benefiting from my social environment and the relationships it provides. And, I find myself still a "closet qualitative researcher," getting my insights from close observation of the world around me and finding research ideas when these observations don't coincide or aren't explainable by the existing theories of our field. I still find it incredible that I am paid well to indulge my curiosity, which is what academic life is about. And, I still don't shy away from a challenge or from controversy. The most recent example has been the furor caused by my article on the state of paradigm development in the organization sciences and its causes and consequences, which has provoked sharp commentary from John Van Maanen, among others. But perhaps that is the role I have settled into and which I find the most interesting—a role in which I say the obvious, except it seldom is as obvious to others, and in which I try to find new ways of understanding organizations, taking ideas and observations from wherever I can.

The journey I have pursued and continue to follow is not one I necessarily recommend to others. It requires one to face setbacks—most of my better theoretical ideas, in particular, have been rejected at least once and often more frequently. It requires one to build new avenues of inquiry rather than following paths that are both safer and have more companions. It requires the ability to not shy away from controversy or confrontation, when necessary, and the wisdom to understand that one is defined by one's detractors as well as one's friends—you should choose both very carefully. And it certainly requires the ability to find colleagues and a network of social support to make all of this possible. In taking the road less traveled, one is well-advised to have good friends to help with the journey. In that sense, academic research, even when such research seeks to be innovative, is inevitably the product of both the individual and the social environment that makes such innovation feasible.

PUBLICATIONS

1971

With H.R. Fogler & T. Deeley. Building and using computerized financial planning simulations. *Simulation and Games, 2*, 213-225.

1972

Size and composition of corporate boards of directors: The organization and its environment. *Administrative Science Quarterly, 17*, 218-228.

Merger as a response to organizational interdependence. *Administrative Science Quarterly, 17*, 382-394.

Interorganizational influence and managerial attitudes. *Academy of Management Journal, 15*, 317-330.

1973

Canonical analysis of the relationship between an organization's environment and managerial attitudes toward subordinates and workers. *Human Relations, 26*, 325-337.

Size, composition and function of hospital boards of directors: A study of organization-environment linkage. *Administrative Science Quarterly, 18*, 349-364.

With H. Leblebici. Executive recruitment and the development of interfirm organizations. *Administrative Science Quarterly, 18*, 449-461.

With H. Leblebici. The effect of competition on some dimensions of organizational structure. *Social Forces, 52*, 268-279.

1974

Administrative regulation and licensing: Social problem or solution? *Social Problems, 21*, 468-479.

Cooptation and the composition of electric utility boards of directors. *Pacific Sociological Review, 17*, 333-363.

With G.R. Salancik. Organizational decision making as a political process: The case of a university budget. *Administrative Science Quarterly, 19*, 135-151.

Some evidence on occupational licensing and occupational incomes. *Social Forces, 53*, 102-111.

With R.E. Miles & C.C. Snow. Organization-environment: Concepts and issues. *Industrial Relations, 13*, 244-264.

With G.R. Salancik & H. Leblebici. Stability and concentration of National Science Foundation funding in sociology, 1964-1971. *American Sociologist, 9*, 194-198.

With G.R. Salancik. The bases and use of power in organizational decision making: The case of a university. *Administrative Science Quarterly, 19*, 453-473.

1975

With J. Dowling. Organizational legitimacy: Social values and organizational behavior. *Pacific Sociological Review, 18*, 122-136.

With G.R. Salancik. Determinants of supervisory behavior: A role set analysis. *Human Relations, 28*, 139-154.

1976

Beyond management and the worker: The institutional function of management. *Academy of Management Review, 1*, 36-46.

With G.R. Salancik & H. Leblebici. The effect of uncertainty on the use of social influence in organizational decision making. *Administrative Science Quarterly, 21*, 227-245.

With A. Leong & K. Strehl. Publication and prestige mobility of university departments in three scientific disciplines. *Sociology of Education, 49*, 212-218.

With P. Nowak. Joint ventures and interorganizational interdependence. *Administrative Science Quarterly, 21*, 398-418.

With H. Aldrich. Environments of organizations. *Annual Review of Sociology, 2*, 79-105.

With P. Nowak. Patterns of joint venture activity: Implications for antitrust policy. *The Antitrust Bulletin, 21*, 315-339.

1977

The ambiguity of leadership. *Academy of Management Review, 2*, 104-112.
Power and resource allocation in organizations. In B.M. Staw & G.R. Salancik (Eds.), *New directions in organizational behavior* (pp. 235-265). Chicago: St. Clair Press.
With A. Leong. Resource allocations in United Funds: An examination of power and dependence. *Social Forces, 55*, 775-790.
With G.R. Salancik. Organizational context and the characteristics and tenure of hospital administrators. *Academy of Management Journal, 20*, 74-88.
With H. Leblebici. Information technology and organizational structure. *Pacific Sociological Review, 20*, 241-261.
With G.R. Salancik. Constraints on administrator discretion: The limited influence of mayors on city budgets. *Urban Affairs Quarterly, 12*, 475-498.
With G.R. Salancik. Who gets power—and how they hold on to it: A strategic-contingency model of power. *Organizational Dynamics, 5*, 3-21.
With A. Leong & K. Strehl. Paradigm development and particularism: Journal publication in three scientific disciplines. *Social Forces, 55*, 938-951.
With G.R. Salancik. Administrator effectiveness: The effects of advocacy and information on resource allocations. *Human Relations, 30*, 641-656.
With G.R. Salancik. An examination of need-satisfaction models of job attitudes. *Administrative Science Quarterly, 22*, 427-456.
With G.R. Salancik. Organizational design: The case for a coalitional model of organizations. *Organizational Dynamics, 6*, 15-29.
The effects of an MBA and socioeconomic origins on business school graduates' salaries. *Journal of Applied Psychology, 62*, 698-705.
Toward an examination of stratification in organizations. *Administrative Science Quarterly, 22*, 553-567.
Usefulness of the concept. In P.S. Goodman, J.M. Pennings, & Associates (Eds.), *New perspectives on organizational effectiveness* (pp. 132-145). San Francisco: Jossey-Bass.

1978

With G.R. Salancik. *The external control of organizations: A resource dependence perspective*. New York: Harper and Row.
With G.R. Salancik & J.P. Kelly. A contingency model of influence in organizational decision making. *Pacific Sociological Review, 21*, 239-256.
Organizational design. Arlington Heights, IL: AHM Publishing.
With G R. Salancik. A social information processing approach to job attitudes and task design. *Administrative Science Quarterly, 23*, 224-253.

The micropolitics of organizations. In M.W. Meyer & Associates (Eds.), *Environments and organizations* (pp. 29-50). San Francisco: Jossey-Bass.
With G.R. Salancik. Uncertainty, secrecy, and the choice of similar others. *Social Psychology, 41*, 246-255.

1980

With J. Lawler. Effects of job alternatives, extrinsic rewards, and behavioral commitment on attitude toward the organization: A field test of the insufficient justification paradigm. *Administrative Science Quarterly, 25*, 38-56.
With J. Ross. Union-nonunion effects on wage and status attainment. *Industrial Relations, 19*, 140-151.
A partial test of the social information processing model of job attitudes. *Human Relations, 33*, 457-476.
With W.L. Moore. Average tenure of academic department heads: The effects of paradigm, size and departmental demography. *Administrative Science Quarterly, 25*, 387-406.
With W.L. Moore. Power in university budgeting: A replication and extension. *Administrative Science Quarterly, 25*, 637-653.
With W.L. Moore. The relationship between departmental power and faculty careers on two campuses: The case for structural effects on faculty salaries. *Research in Higher Education, 13*, 291-306.
With G.R. Salancik. Effects of ownership and performance on executive tenure in U.S. corporations. *Academy of Management Journal, 23*, 653-664.

1981

Power in organizations. Marshfield, MA: Pitman.
Management as symbolic action: The creation and maintenance of organizational paradigms. In B.M. Staw & L.L. Cummings (Eds.), *Research in organizational behavior* (Vol. 3, pp. 1-52). Greenwich, CT: JAI Press.
With J. Ross. Unionization and female wage and status attainment. *Industrial Relations, 20*, 179-185.
Four laws of organizational research. In A.H. Van de Ven & W.F. Joyce (Eds.), *Perspectives on organization design and behavior* (pp. 409-418). New York: Wiley-Interscience.
Some consequences of organizational demography: Potential impacts of an aging work force on formal organizations. In S.B. Kiesler, J.N. Morgan, & V.K. Oppenheimer (Eds.), *Aging: Social change* (pp. 291-329). New York: Academic Press.
With J. Ross. Unionization and income inequality. *Industrial Relations, 20*, 271-285.

1982

Organizations and organization theory. Marshfield, MA: Pitman.
With J. Ross. The effects of marriage and a working wife on occupational and wage attainment. *Administrative Science Quarterly, 27*, 66-80.

1983

Organizational demography. In L.L. Cummings & B.M. Staw (Eds.), *Research in organizational behavior* (Vol. 5, pp. 299-357). Greenwich, CT: JAI Press.
With M.L. Markus. Power and the design and implementation of accounting and control systems. *Accounting Organizations and Society, 8*, 205-218.
With B.E. McCain & C.A. O'Reilly. The effects of departmental demography on turnover: The case of a university. *Academy of Management Journal, 26*, 626-641.

1984

With Y. Cohen. Employment practices in the dual economy. *Industrial Relations, 23*, 58-72.
With W.G. Wagner & C.A. O'Reilly. Organizational demography and turnover in top-management groups. *Administrative Science Quarterly, 29*, 74-92.
With Y. Cohen. Determinants of internal labor markets in organizations. *Administrative Science Quarterly, 29*, 550-572.

1985

Organizations and organization theory. In G. Lindzey & E. Aronson (Eds.), *Handbook of social psychology* (Vol. 1, pp. 379-440, 3rd ed.). New York: Random House.
Organizational demography: Implications for management. *California Management Review, 28*, 67-81.

1986

With Y. Cohen. Organizational hiring standards. *Administrative Science Quarterly, 31*, 1-24.
With A. Davis-Blake. Administrative succession and organizational performance: How administrator experience mediates the succession effect. *Academy of Management Journal, 29*, 72-83.

1987

A resource dependence perspective on intercorporate relations. In M.S. Mizruchi & M. Schwartz (Eds.), *Intercorporate Relations* (pp. 25-55). Cambridge: Cambridge University Press.

With A. Davis-Blake. The effect of the proportion of women on salaries: The case of college administrators. *Administrative Science Quarterly, 32*, 1-24.

Bringing the environment back in: The social context of business strategy. In D.J. Teece (Ed.), *The competitive challenge* (pp. 119-135). Cambridge, MA: Ballinger.

With C.A. O'Reilly. Organizational demography and turnover among nurses. *Industrial Relations, 26*, 158-173.

With A. Davis-Blake. Understanding organizational wage structures: A resource dependence approach. *Academy of Management Journal, 30*, 437-455.

1988

With J. Baron. Taking the workers back out: Recent trends in the structuring of employment. In B.M. Staw & L.L. Cummings (Eds.), *Research in organizational behavior* (Vol. 10, pp. 257-303). Greenwich, CT: JAI Press.

With J. Ross. The compensation of college and university presidents. *Research in Higher Education*, pp. 79-91.

With N. Langton. Wage inequality and the organization of work: The case of academic departments. *Administrative Science Quarterly, 33*, 588-606.

1989

With A. Davis-Blake. Just a mirage: The search for dispositional effects in organizational research. *Academy of Management Review, 14*, 385-400.

The politics of careers. In M.B. Arthur, D.T. Hall, & B.S. Lawrence (Eds.), *Handbook of career theory* (pp. 380-396). New York: Cambridge University Press.

1990

With J. Ross. Gender-based wage differences: The effects of organizational context. *Work and Occupations, 17*, 55-78.

With A. Davis-Blake. Determinants of salary dispersion in organizations. *Industrial Relations, 29*, 38-57.

Incentives in organizations: The importance of social relations. In O.E. Williamson (Ed.), *Organization theory: From Chester Barnard to the present and beyond* (pp. 72-97). New York: Oxford University Press.

With A. Konrad. Do you get what you deserve? Factors affecting the relationship between productivity and pay. *Administrative Science Quarterly, 35*, 258-285.

With A. Davis-Blake. Unions and job satisfaction: An alternative view. *Work and Occupations, 17*, 259-283.

1991

With A. Konrad. Understanding the hiring of women and minorities in educational institutions. *Sociology of Education, 64*, 141-157.

With A. Konrad. The effects of individual power on earnings. *Work and Occupations, 18*, 385-414.

Organization theory and structural perspectives on management. *Journal of Management, 17*, 789-803.

1992

Managing with power: Politics and influence in organizations. Boston: Harvard Business School Press.

With A. Davis-Blake. Salary dispersion, location in the salary distribution, and turnover among college administrators. *Industrial and Labor Relations Review, 45*, 753-763.

1993

With N. Langton. The effect of wage dispersion on satisfaction, productivity, and working collaboratively: Evidence from college and university faculty. *Administrative Science Quarterly, 38*, 382-407.

Barriers to the advance of organizational science: Paradigm development as a dependent variable. *Academy of Management Review, 18*, 599-620.

1994

Competitive advantage through people: Unleashing the power of the work force. Boston: Harvard Business School Press.

The costs of legalization: The hidden dangers of increasingly formalized control. In S.B. Sitkin & R.J. Bies (Eds.), *The legalistic organization* (pp. 329-346). Thousand Oaks, CA: Sage.

With N. Langton. Paying the professor: Sources of salary variation in academic labor markets. *American Sociological Review, 59*, 236-256.

With J. Baron. The social psychology of organizations and inequality. *Social Psychology Quarterly, 57*, 190-209.

1995

With A. Davis-Blake & D.J. Julius. AA officer salaries and managerial diversity: Efficiency wages or status? *Industrial Relations, 34*, 73-94.

NOTES

1. V.H. Vroom. (1964). *Work and motivation*. New York: Wiley.
2. A.G. Weinstein & V. Srinivasan. (1974). Predicting managerial success of Master of Business Administration (MBA) graduates. *Journal of Applied Psychology, 59*, 207-212.
3. J. P. Kotter. (1982). *The general managers*. New York: Free Press.
4. C. Faucheux & K. D. Mackenzie. (1966). Task dependency of organizational centrality: Its behavioral consequences. *Journal of Experimental Social Psychology, 2*, 361-375; K.D. Mackenzie. (1978). *Organizational structures*. Arlington Heights, IL: AHM Publishing.
5. J.D. Thompson. (1967). *Organizations in action*. New York: McGraw-Hill.
6. D. Katz & R.L. Kahn. (1968). *The social psychology of organizations*. New York: Wiley.
7. D.S. Pugh, D.J. Hickson, C.R. Hinings, & C. Turner. (1968). Dimensions of organization structure. *Administrative Science Quarterly, 13*, 65-105.
8. K.E. Weick. (1969). *The social psychology of organizing*. Reading, MA: Addison-Wesley.
9. L.W. Porter & E.E. Lawler. (1968). *Managerial attitudes and performance*. Homewood, IL: Irwin-Dorsey.
10. R.I. Sutton. (in press). The virtues of closet qualitative research. *Organization Science*.
11. E.J. Webb, D.T. Campbell, R.D. Schwartz, & L. Sechrest. (1966). *Unobtrusive measures*. Chicago: Rand McNally.
12. J. French, Jr., & B. Raven. (1968). The bases of social power. In D. Cartwright & A. Zander (Eds.), *Group Dynamics* [*pp. 259-269, 3rd ed.*]. *New York: Harper and Row.*
13. *J. Lodahl & G. Gordon.* [*1972*]. *The structure of scientific fields and the functioning of university graduate departments. American Sociological Review, 37*, 57-72.
14. See, for example, S. Cole. (1983). The hierarchy of the sciences? *American Journal of Sociology, 89*, 111-139; S. Cole, J.R. Cole, & G.A. Simon. (1981). Chance and consensus in peer review. *Science, 214*, 881-886.
15. R.S. Burt. (1983). *Corporate profits and cooptation*. New York: Academic Press.
16. C.P. Alderfer. (1977). A critique of Salancik and Pfeffer's examination of need-satisfaction theories. *Administrative Science Quarterly, 22*, 658-669.

A Taste for Innovation

DEREK S. PUGH

Being 65 years old, I retire at the end of this month (September 1995)—the British university system being old fashioned enough (or maybe forward thinking enough) to have a mandatory retirement age. I am not sure what retirement would mean, other than going freelance. I certainly have commitments for the next few years to two ongoing research programs, and the list of books that I am contracted to write and series to edit appears, if anything, to be growing. Since "someone who thinks retirement means going freelance" is as good a definition of a workaholic as any, and I have always insisted that I am not one (to my family's cries of derision it must be admitted), *things will change*, even though I am not sure just how or when. This is therefore an appropriate occasion for me to reflect on my professional career.

EARLY YEARS

I was born on 31 August 1930 in London. My grandparents were Jewish immigrants who came to London at the turn of the century. My paternal grandfather's name was PUCH (Russian for "down;" i.e., as stuffed into a duvet) but the name had slid over to PUGH by the time that my birth was registered. I am tickled that when I check out of a hotel in Continental Europe or Israel, the process often goes into reverse and I get the bill in my grandfather's name. I derived immense benefit from this background in the high aspirations

Management Laureates, Volume 4, pages 235-276.

ISBN: 1-55938-730-0

communicated to me, and in the high value put on education. My parents were comfortably middle class when I was born but suffered soon afterwards in the Depression. But there was never any suggestion that I should not continue to study for as long as necessary, and an academic career was considered to be of the highest status.

At school I was originally interested in science, with medicine as the intended profession. My adolescent rebellion consisted of reading a very great deal, and rejecting this route. I was interested in modern ideas of education and was much taken with A.S. Neill's books about his school's liberal practices, as influenced by the American educator, Homer Lane. At those schools lessons were voluntary, and pupils were allowed to vote on decisions concerned with the schools' running. Neill's fascinating psychological explanations as to why this worked to produce mature educated children, even when he took rebellious or disturbed children from other schools, set me on to reading psychology. I read about Freud, Jung, Adler, Kohler and enrolled in an adult education evening class to hear of Pavlov and Piaget. I decided that I wanted to study this, then rather innovative, subject of psychology at university.

EDINBURGH PSYCHOLOGY

I was fortunate to be able to study at the University of Edinburgh, with a grant from the London County Council. Traditional Scottish universities are different from English ones. With a four year degree framework and a modular structure, they are much closer to the American pattern. At the time I was a student (1949-1953), the Edinburgh Psychology Department was offering by far the most thorough education in all aspects of psychology in Britain. In addition to the wide range of lecture courses offered, the opportunities for practical experience were immense. I ran animal experiments (pigeons, I'm glad to say; our learning theory man was a Skinnerian), gave intelligence tests to schoolchildren and to paraplegics (Terman-Merrill and Koh's Blocks respectively, if I remember rightly) and administered and interpreted projective techniques (Rorshach, TAT, Szondi).

Edinburgh was then a major center for printing and publishing. The selection of boys as printers' apprentices (no girls of course, which shows how long ago this was) was undertaken by the Psychology Department. Honors students were involved in administering the tests, conducting the interviews, writing evaluations and participating in selection decisions. This really caught my interest; psychologists doing a real job in a practical situation that I could relate to. I found psychiatric hospitals too depressing, and I am afraid that I got a bit bored with the fact that theories of behavior were supposed to be furthered through the study of white rats. That schools and firms could make use of applied psychologists was much more interesting.

Conceptually the range of studies was wide. In addition to psychology I took courses in mathematics, social anthropology and political economy (as economics was called in my time). The University regarded the fledgling science of psychology as having so recently moved out from under the wing of philosophy that we were all required to take several courses in logic and metaphysics. I took to philosophy immediately. Thus I had a good grounding in epistemology, and, as is appropriate for one studying on the same benches as David Hume, had my period as a solipsist.

In due course, I came to the conclusion that, whatever view I took on the philosophical issues as such, if I was going to pursue any substantive topic I would have to assume a realist, determinist approach to analysis. Otherwise I would be condemned to spend my time on the metaphysics without getting round to the physics, or, in my case to the social psychology. This is still my view, which is why I refer to myself as an "unreconstructed positivist."

In 1951, in the midst of this welter of ideas, I met Natalie A. Gorovitz. She was a sociology student at the London School of Economics, and introduced me to that discipline. Sociology was not, at that time, taught in Edinburgh and I learned from her about Hobhouse, Durkheim and Talcott Parsons. She it was who gave me my first sight of Roethlisberger and Dixon's *Management and the Worker*; expensive American books being not then easily available in Britain. That was a fascinating revelation—the *social* psychology of industry (as J.A.C. Brown was later to put it in the title of a book). Her interests too developed into the industrial sphere as a vocational guidance counselor. What was there to do? "Reader, I married her!"

During my career I have learnt from many people. My wife has been a fellow professional and a constant source of ideas over the years, and I have learnt much from many colleagues. But at the early stages in my career I worked with four senior colleagues whom I came to regard as my "professional father figures." The relationship with some of them was not always easy (there is, after all, a strong Oedipal element in all Western parent-children relationships), but they played a major part in my intellectual formation. The first of these, and the only one still living, was Boris Semeonoff of the Edinburgh Psychology Department. He is, I am glad to say, still going strong in his eighties, having recently been elected an Honorary Fellow of the British Psychological Society in recognition of his contributions to the discipline over more than sixty years. What I took from Boris was breadth of vision. He was both the leading statistical expert in the department and the leading exponent of the subtle individual, and often psychoanalytical, analysis involved in projective techniques. In a highly fractionated discipline and department, this was distinctive.

Of course, at the beginning I did not really understand that the relationships between nomothetic and ideographic study are very subtle. I cheerfully carried out, as my MA project, a statistical study of the Rorschach configurations of children of low arithmetical achievement compared with their linguistic attainment. I shudder now at the crassness of it, but it did mean that I got my first publication at the age of 23. I had spotted that you can get an interesting paper out of negative results, by comparing your data with somebody else's work. The report appeared in 1954 in the *Journal of Projective Techniques*, right next to an article by Rorschach himself, no less,—not a communication from the spirit world but a newly discovered, posthumous paper.

SOCIAL SCIENCES RESEARCH CENTER, EDINBURGH

But by then I was working with my second professional father figure and was growing up very quickly both intellectually and organizationally. After I graduated I was invited by Roderick M. McKenzie ("Mac" to everybody) to be his research assistant on a newly funded project. After the Second World War, the poor financial state of Europe benefited from the U.S. Government's Marshall Aid program. In the early 1950s there was a follow-up program in which aid in dollars (i.e., hard currency) was given by America on condition that an equivalent amount of money in the country's own currency was spent on improving its industrial productivity. This "Counterpart Aid" allowed a number of industrial social science research projects to be set up in Britain and, for my generation, provided a large increase in the number of research job opportunities.

I therefore finished the old academic year as a student and started the new one as a Research Assistant in the same department. It is true that RAs are the lowest form of academic life, but they *are* members of staff and a whole new world opened for me. As a student I had known in a general way that there were disagreements in the department and some lecturers were not too friendly with others. But I had no idea of the degree of contempt, conflict and hatred involved in the cross-currents until I joined the staff. As McKenzie's assistant, I was taken under his wing and inevitably saw the issues primarily from his point of view, but I was also impressed by his organizational knowledge and subtlety in understanding and explaining the motivations of the other players.

Although Mac and I were members of the psychology department, we were seconded to the Social Sciences Research Centre. The reason why there was no sociology taught at that time in the normal undergraduate program at Edinburgh (it was only taught in the Social Studies department as part of the professional training of social workers) was that the government grant had been used instead to found this interdisciplinary Centre. All the academic members had been seconded from social science departments with the idea of developing a more integrated research approach which it was hoped would get us further in understanding social phenomena. Those seconded were very impressive: they included Tom Burns (the sociologist from the department of Social Studies, who was soon to be joined on another Counterpart Aid project by the psychologist, George Stalker), Michael Banton (from Social Anthropology, later a leading sociologist of contemporary Britain) and Hilde Behrend (from Commerce, a leading academic in industrial relations). Irving Goffman had been a visiting scholar (and my tutor in social anthropology), and the first edition of *The Presentation of Self in Everyday Life* was published as a Centre monograph.

As a beginning academic, I bought entirely the innovative message that interdisciplinary integration was the way forward. Looking back now, I see the Centre as a heroic, yet doomed, enterprise. There was no Director—the idea appeared to be that it would function as a sort of intellectual *kibbutz*. Tom Burns was already the biggest name, and was later to become much bigger with the *Management of Innovation* book. He was sometimes regarded by outsiders as the leader, but he did not wish to play that role internally. And certainly several others would have resisted that.

The Centre is long gone; Edinburgh has a fine Sociology department of which Tom Burns became the first Professor and Head. But I retain something very important from my experience there. It is a great skepticism about academic disciplines; in particular about the arbitrariness of their boundaries. I can easily be provoked into maintaining that they are merely the restrictive work practices of academics. What matters is that the subject of study should be illuminated in as many different ways, and with as many conceptual research schemes, as possible. And while collaboration across disciplines is good and should be encouraged, real interdisciplinary integration takes place in the researcher's head. The clearest way in which this view has stayed with me is that in my academic career I have eschewed traditional disciplines. In the succeeding four decades I held posts in Social Medicine, Human Relations, Industrial Administration, Organizational Behavior, Systems and International Management.

The project that Mac and I carried out at the Centre concerned the problems of inspection in British industry. Although we did look at some problems of selection and training in the traditional industrial psychology mode, the focus of the study was a post-Hawthorne social interaction approach. On my part, this entailed a reading of the early Human Relations literature (B.B. Gardner, I remember, appeared very wise and, when we were feeling depressed, seemed to have said it all). I also tackled the management literature and found, as you would expect, that F.W. Taylor had some forceful and shrewd things to say about inspection.

During this project I learned many things. From Mac I learned about being sensitive to an individual's motivation, the key importance of status and how it is manipulated, and how interpersonal control works. For myself I had already begun to take seriously the notion of organizational structure. Tom

Burns talked about Weber and, after I discovered that this was not the psychologist who linked the intensity of the stimulus to intensity of the sensation (i.e., the Weber-Fechner Law), nor the musician who wrote "Invitation to the Dance" the only Webers I knew, I found out a bit more about Max Weber. (I feel the need for a Wagnerian *leitmotiv* at this point!) The aspect of the study that I wrote up for my MSc dissertation was concerned with the impact on the inspection-production relationship of the differences in their positions in the organizational structures of the Chief Inspectors of three subsidiaries of one firm.

This aspect of the project was eventually published in a paper in the *Journal of Management Studies* about ten years later. This brings me to another thing I learned from Mac, in the "how not to do it" vein. Research is not carried out unless the results are published. Mac had a problem with publication, which is why he is not nearly as well known as he should have been. He was the classic perfectionist; always putting off publication in order to improve the work. But he was also always finding ways in which teaching—and he was a fine teacher whose students regularly considered him the best they had experienced—would take priority over writing. Since he was the designer and leader of the inspection project, when I left I felt it right to allow him to publish overall from it. But after ten years when little had appeared relative to what we had investigated, I decided to publish anyway and my later articles appeared in 1966.

There is a sequel. In his sixty-fifth year on the eve of his retirement, Mac died after a painful cancer, bravely borne. During his illness he talked with friends about an unpublished manuscript on which he had worked after I had left. It was clear that the work meant a lot to him. After his death, Hilde Behrend arranged for it to be considered for publication and asked me to edit it. I found the manuscript fascinating; full of penetrating insights based on the psychological understanding, concentrated detailed analysis, and sheer practical common-sense that characterized Mac's approach. The volume was published in 1989—but that was too late.

During this period I also gently began my teaching career. Since, like the rest of us, I have suffered from so much inadequate teaching in my time, striving to be an effective teacher has always been important for me. So are innovative teaching methods, and I have tried to contribute to their development. Indeed my first teaching task in the mid-1950s was innovative in that I had heard of no other institution where it occurred. Because no entry knowledge of statistics was required for the first-year psychology course, many students found the weekly statistics lecture difficult to follow on merely one hearing. A second weekly lecture in which the same topics were covered was therefore set up, and about half the class attended it. Giving this "echo lecture" was my first teaching job. I had to put on an academic gown (borrowed), enter the main lecture theater, and discourse to the assembled multitude (well, about a hundred). Being a ham, I loved it of course, and was sorry that I was allowed to do it for only one year.

EDINBURGH SOCIAL MEDICINE

My next job was in the Department of Public Health and Social Medicine, where I worked on a project concerned with sickness absence as a social phenomenon. The project was led by Cecil Gordon, a social biologist who during the Second World War played a leading part in the development of Operational Research in the Royal Air Force, and later in the foundation of the British Operational Research Society. It had been designed with Roy Emerson, whom I succeeded. I came in at the data analysis stage, my first involvement in a large scale statistical survey. As I am fond of pointing out, I can't program a computer—by the time they came in I could loftily require others to do that for me—but I used to be able to plug up a Hollerith board! When I came to know him, Cecil Gordon had suffered a number of debilitating physical and mental illnesses. Although he was a general benign presence presiding over the research activity, he could not, in fact, make an intellectual contribution. Although I was the junior member of the team, I had to become responsible for organizing and writing up the material for academic publication. Intellectually I recognized this after about two weeks; but emotionally it took me nearly three months to accept that nothing was going to happen unless I did it. I then buckled down to the analysis and writing up, which meant that my name was on three articles from about one year's work.

At this time, as an applied psychologist, I was asked by the Dean of the Faculty of Medicine to evaluate the adequacy of the selection procedures into the Medical School. The cynics had long said that the sons of doctors, particularly if their fathers were Edinburgh graduates, were admitted on poorer school (i.e. high school) leaving attainment than others. The orthodox view held that this was not so. I found that it was so. Then an interesting thing happened: the establishment's denial changed to justification. The sons understood better what was involved in the job, and indeed the life, of a medical practitioner; they would be more committed to last the long and grueling course; their fathers were in a better position to help them to become effective more quickly; and so on. Nothing changed.

BIRMINGHAM COLLEGE OF TECHNOLOGY

It was Natalie who brought to my attention in 1957 the advertisement for a Lecturer in Human Relations at the Birmingham College of Technology, and urged me to apply. This was probably the most important single decision of my professional life. The College had been founded as a technical college concentrating on part-time sub-degree education, but it had recently been designated to be developed into a College of Advanced Technology. This meant that it would be able to present "degree equivalent" courses and, most importantly, would be expected to develop its research activities. As an earnest of this intention a number of "Reader" posts were established. In the British system, these are senior posts with the emphasis on research activity—not being universities they could not establish professorships. In the "Birmingham Tech" Department of Industrial Administration where the management education took place (in Britain in those days, "business schools" meant secretarial colleges) an internal promotion to the newly established Readership meant that a lectureship became available, and I was appointed.

When I decided to go to this strange low status institution, the reactions from my Edinburgh colleagues were very mixed. While some were intrigued that applied social science was wanted there, several warned me that I would never get back into a university if I went. The fact that it was a permanent post, with an increased salary was a plus, of course. But the move from Edinburgh to Birmingham was a minus which would take me a considerable time to adjust to. But the more I found out about the proposed developments, the more interested I became in the potential for change and innovation.

On arrival, I was immediately thrown in at the deep end into a teaching schedule which required me to teach for up to 20 hours a week, including two evenings and Saturday mornings, for which I had one weekday off. But of course, with Sod's Law in full operation, that was a day on which I had to teach in the evening. That was the bad news. The good news was that all the students were in jobs. They came to the college in the evenings, or on day release arranged through their employers. This meant that courses were repeated in parallel. For example, you could do the Certificate in Workshop Supervision on Tuesday afternoons and Friday evenings, or on Wednesday evenings and Saturday mornings. Staff were all required to repeat their classes several times in one week. This is an excellent way for new lecturers to learn their trade—*if* they get help in preparation, and feedback on their performance.

I was extremely fortunate in receiving this from my third professional father-figure, John Munro Fraser. Munro, as he was known, was very different from Mac. They knew of each other but I don't think they ever met. Had they done so I am sure they would soon have cordially disliked each other—they did have in common that they were both good haters. Mac was subtle, wanting to build on intuition, wondering about the reasons behind the reasons; Munro was direct, go-getting, wanting an analysis which would give a formula to wrap the matter up. They were both Scotsmen, but Mac fitted my stereotype of an Italian, Munro of an American. Mac found it hard to get academic material to the stage of publication; Munro published a steady stream of books. Mac had developed a system of classification of jobs for use in vocational counselling which was ten years ahead of its time. He published a couple of papers on it, but never managed to complete for publication the definitive handbook that the system required to become properly established. Munro published several workbooks on his "Five-fold Grading" selection system which was a well-known rival to Alec Rodger's "Seven Point Plan."

Munro had designed the courses in Human Relations that we in the section taught. He specified the contents of each lecture. The subsequent discussion topics were designed to enable the participants, who had good practical industrial experience as supervisors, technical specialists, junior managers, or shop stewards, to explore the ideas and test them against their experience. Having to conduct the same class four times in one week, certainly polished my presentation skills. The hardest task, I found, was to be able to remember whether the joke I was about to tell had been used yesterday, (in which case this was a different class and I could use it again), or last week (in which case it was the same class and had to be avoided). This assembly line approach did not apply to certain specialist courses. I developed and presented my own course on staff selection, including Munro's methods, of course, but going wider. This was a very welcome increase in autonomy. I don't suppose this experience would suit everyone, but I must say it gave me a good grasp of the skills of classroom teaching.

In addition to my Human Relations teaching, I had to contribute to the course on Management Principles and Practice. This was the final course of the Diploma in Industrial Administration and was intended to be the jewel in our crown. It consisted of a series of case studies which attempted to pull together all the previous teaching. Each course was taught on a visiting basis by a senior business manager from the Birmingham area, with any required academic input given by a member of staff. In my first year, I was allocated to work with a very unusual man who is my fourth professional father figure. Joe Hunt (later Sir Joseph Hunt) was the managing director of a very successful hi-tech automation company "Hymatic Engineering." He ran his firm in what Tom Burns had called an organic (later organismic) way. This is impressive enough, but Joe was a successful top manager who was also a natural born teacher. He just knew, without as far as I know having had any teacher training, that in this field, the important task of a teacher is not to give the right answer, but to ask the right question. He was naturally skilled in doing this, and his students and myself, as his junior colleague, learned a great deal from him of the subtleties of management. He was open in thought too, again unusual in a successful senior manager, and learned from us. I regard it as a considerable accolade that he asked for me to be his co-tutor on this course in the two subsequent years that I was available.

In my second year, Norman R.F. Maier came for a month as a visiting teacher. He introduced his role playing exercises, and this was a big eye-opener. It is still the only teaching methodology that I know which actually benefits from larger class sizes through comparisons between the smaller sub-groups.

I began to use some of his role plays, and developed a couple of my own. I used Norman's exercise, "The Change of Work Procedure," regularly for the next twenty years. It is an excellent introduction to interpersonal leadership skills.

During this year too, I began working on an innovative project with two colleagues not in the Human Relations section. As can be imagined, this was an unusual thing to do in a mechanistic set-up such as ours, but it grew out of my commitment to integration in teaching. Bill Williams (an economist) and John Fairhead (from the business communications section) and I started work to develop our version of a non-computer business game based on G.R. Andlinger's 1954 paper in the *Harvard Business Review*. Over the succeeding years we developed, with other colleagues, many business exercises which we later collected and published as *Exercises in Business Decisions* in 1965.

Also at this time I began my first experience of distance teaching (another *leitmotiv* here). Tom Wylie, a former trade union official who headed our Industrial Relations section, suggested me as the correspondence tutor in social psychology for Ruskin College, Oxford. Ruskin was the college that prepared trade unionists with no qualifications for entry into the university proper. But it also ran correspondence courses for shop stewards and other workplace representatives. They had to read material sent to them, and send me essays that I commented upon. No marking of course; this was true education for development. I enjoyed it, and did a six-year stint, until the system changed.

After three years of concentration on teaching at this intensity, I felt that I had had enough and was more than ready for a change when the opportunity came.

THE INDUSTRIAL ADMINISTRATION RESEARCH UNIT

The change came in 1960 with the appointment of a new head of the Department of Industrial Administration. The previous head, David Bramley, had left when it became clear that the designation as a College of Advanced Technology was going to lead the institution to establish Bachelor's and even Master's degrees. He felt this was an academic diversion from what the College should be doing, and so returned to an industrial post. The new person appointed was a surprise: Tom Lupton, a social anthropologist from the department at Manchester. The choice of a businessman or an engineer was expected, and Tom's appointment heralded that the Board of Governors of the College was taking very seriously the intended development in the academic standing of the organization.

Tom Lupton had brought with him a large Government grant for the study of shop floor behavior in British factories. But since he had been appointed as the head of the largest management studies department in the country, he found, inevitably, that he had no time to launch the research. Nor did any of his research colleagues in Manchester transfer with him to Birmingham. This was my opportunity. I offered to be seconded from my lectureship to work on the research, and Tom agreed enthusiastically. There was also a College Research Fellowship available and David Hickson was appointed to it.

I don't remember when I first met David—though it was as momentous for me as Oliver Hardy first clapping eyes on Stan Laurel. But I still do remember very clearly the look on his face when I told him that I was going to leave my lectureship to do research full time, and that we were going to set up a research unit and intended to appoint other researchers. *He had not been told.* He had thought that he was going to be a lone researcher working under the general supervision of Tom Lupton, but he readily agreed to join the group. Thirty-five years later we still regularly work on collaborative projects. The first book on which we collaborated was dedicated to our parents (and our professional father-figures); the latest to our grandchildren.

The Industrial Administration (I.A.) Research Unit was set up on 1st January 1961, with a Senior Research Fellow (myself) and a Research Fellow (David). We appointed two Research Assistants: Bob Hinings, a sociologist, and Graham Harding, a psychologist. Bob thus started his long association with the work, and his subsequent major contributions to the field of organizational analysis.

It soon became clear that Graham Harding's prime interest was in experimental psychology. He had applied for the job because he was from Birmingham and wanted to return to live and work there. His selection was due to a failure of imagination on my part—even though I was supposed to be a specialist in selection procedures! I had considered academically relevant posts all over the country, and the idea that someone would look for a job geographically rather than professionally was strange. Graham could help us with his down-to-earth approach and his knowledge of local industry—he was more streetwise than the rest of us. But basically he was not interested enough conceptually to participate fully. When, after a year, the opportunity of the Wilmot Breeden Research Fellowship came up, I encouraged him to apply and propose an experimental EEG project for which the equipment was available. He was appointed to the post, and never looked back. He is now a leading physiological psychologist. I saw him on TV recently, debating whether a full knowledge of the workings of the brain will ever be able to "explain" consciousness. He is still at the institution as the Professor of Neuro-ophthalmology.

So the group got started. We retired to our researchers' ivory tower to review the field and plan our program. The "ivory tower" was, in fact, the basement of a nearby slum. But conceptually it was an ivory tower, because we had the great privilege of being allowed to get on with our research. No one else in the Industrial Administration Department was particularly interested in what we were doing, since this was solely a teaching department. Tom Lupton was the exception, of course, and he provided the crucial protection of authority, and much encouragement. We spent a year surveying previous work and hammering out our conceptual framework of context, organization, group, and individual levels of study and their relationships. Everyone participated in all aspects of designing the program of work and carrying out pilot interviews, and this had a great integrating effect.

After Graham Harding went to other work, Bob Hinings moved to a teaching post in the neighboring University of Birmingham, but continued his contribution from his new base. This allowed the recruitment of the "second generation" of researchers. It was rather different for them. Since the conceptual framework and research strategy had been worked out, their contribution was to the operationalization of the concepts at the context and organizational levels, and data collection and analysis. Important as this is, it brings with it the inevitable feeling of working on "somebody's else's research" and they did not develop a long-term commitment to this academic field. Chris Turner works in the field of social services provision, Theo Nichols is a powerful Marxist analyst of modern industry, and Keith Macdonald publishes on the sociology of professions.

I realized that we would need some methodological help, even though I was by now a Fellow of the Royal Statistical Society and held their certificate. I also realized that we did not need a professional statistician (who would only tell us what we could not do) but a methodologically sophisticated social scientist. We were extremely fortunate in interesting a psychologist, Phil Levy of the University of Birmingham, in working with us. He made a major contribution to structuring our analyses, which were way ahead of anything else being done in the field at that time. I have always felt that my highest methodological accolade was not my Fellowship, but the fact that Phil said to me "I enjoy explaining these ideas to you. You get their implications."

THE UNIVERSITY OF ASTON IN BIRMINGHAM

The 1960s was a great time for those of us in higher education—I know I'm getting old when I have attacks of nostalgia and start pitying those of my colleagues who were not around then to experience them. All institutions were expanding, and new ones being formed, as the Government (under both political parties) made ever more resources available. The Birmingham College of Technology became the College of Advanced Technology which, in turn, became the University of Aston in Birmingham. A phenomenal progress—all in a matter of six years or so. Undergraduate degrees were established, Masters' degrees were being designed, the expansion of the Industrial Administration Department into a Faculty of Behavioural Sciences was put in train.

One change I had not bargained for, since it went against my predictions: a difficulty about being a positivist is that the data can bite back! I had always predicted that academics who become heads of departments, with the increased salary, status and power involved, would not give them up just because they preferred to do research. Well, Tom Lupton proved me wrong. He gave up his Aston post to go to a Research Chair at Leeds. It is true that he went from there to the Manchester Business School and, in due course, became its Director (which is half a vote for my hypothesis), but he did demonstrate a commitment to personal research work that is rare among British heads of schools.

By 1965 the external grant was coming to an end. In those days one of the conditions laid down by the Government on providing support such as this, was that the institution would continue to fund the work from its own resources if, at the end of the original grant, the research was still found to be important and timely. With the inevitable bureaucratic complexities involved in such a decision, it was literally within two weeks of the official date of the grant ending that the decision was announced. By this time those on short term contracts had left; in an expanding social science market they had obtained permanent teaching appointments elsewhere. Only David Hickson, who had obtained a Research Lectureship, and I remained.

But, glory be, the answer was "Yes;" the University agreed to continue funding the work. This sort of major support was infrequent even in the 1960s and, for me, it underlines how appropriate it is that the research has become internationally known as "the Aston studies." The decision opened the way for the third generation of Aston researchers. Kerr Inkson, Roy Payne, and Diana Pheysey were paid for by the university. John Child was funded from an additional Government grant that I obtained.

We now entered a new phase of the work: new projects were designed (e.g the wider national organizational study), the previous conceptual frameworks were extended (e.g. to group level work), new concepts were developed and operationalized, new data were collected and analyzed. All the members of this generation were thus involved in all stages of the research, and it is interesting that they all continue to do leading work in the field of organizational studies: Kerr Inkson in Auckland on corporate excellence, Roy Payne in Sheffield on stress, and John Child in Cambridge on management in China. Diana Pheysey worked on corporate change at Aston until her retirement, and I am most touched and grateful that her innovative book on *Organizational Cultures: Types and Transformations* (Routledge, 1993) was dedicated to me.

The Unit was always very fortunate in the high level of contributions it received from its support staff—only one bad selection decision there, over which we will hastily draw a veil. Our information, data and interviewing assistants did us extremely proud. I like to think that we contributed to their development too, in that Cindy Fazey, Rita Austin and Will McQuillan—all of whom started with us as non-graduates—in due course became academics in their own right with lectureships at universities. Patricia Clark has stayed at Aston for three decades in research information. Our secretary, Ruth Goodkin, was a phenomenon in coping so unflappably with a whole bunch of demanding academics.

THE ASSOCIATION OF TEACHERS OF MANAGEMENT, *WRITERS ON ORGANIZATIONS*, AND PENGUIN BOOKS

The history of the Aston studies has been written up several times now, and is beginning to take on a myth-like quality even for those of us who experienced it. I like to be as realistic as possible in describing the tensions and failures as well as the commitment and the successes. Roy Payne once paid me the compliment of introducing me as the most participative manager that he knew. But, as I pointed out, I had no choice: that's how it had to be. For example, in the early years I was the leader of the group, the most experienced researcher, the only member in a permanent post, and, at the time of the second generation, the only psychologist. It does not need a great deal of psychoanalytical insight to see that I would, on occasions, become the focus of the hostility of junior members when the frustrations became too great. There was an inevitable tension between the impact of their short term contracts and my long term view of the objectives of the program. This result just had to be lived with, but it made me grateful for the fact that I was involved in The Association of Teachers of Management (ATM).

The ATM was distinctive to Britain in that it contained professionals from a wide range of institutions concerned with management education. In particular, it was a forum in which academics from colleges and management development officers from industry could meet together to discuss professional issues. It started in 1960: Tom Lupton became the first Chairman, Frank Heller the first Secretary. I was a founder member and accepted the post of editor of the *Newsletter*. This was a mimeographed sheet produced in the Department; my first issue was, in fact, typed up by Bob Hinings. During the next six years I built it up into a larger, professionally produced *ATM Bulletin*, which reflected the growth of activity in the field. This task took some of my time, and that of our information assistant Cindy Fazey, to occasional rumblings from the populace about lost research time. But the activity was very important to me psychologically, in that it engaged me in a professional activity away from the group. I needed the bolt hole. I was also hammering out my view of the nature of our subject and published an article in the *Psychological Bulletin* in 1966 on "Modern Organization Theory," which generated the largest number of requests for offprints, and appeared in the greatest number of anthologies of any of my papers.

It was through the ATM that I came to know Morris Brodie of the Administrative Staff College at Henley. The College had produced a booklet to hand out to its students introducing them to some of the management writings. Morris asked me to prepare a new edition of it. I recruited David Hickson and Bob Hinings and when we had worked out what we wanted to do, it was no longer a booklet, but a book. Thus began *Writers on Organizations*: a set of summaries of the work of leading writers in the field. The first edition in 1964 was a hardback, but the second and subsequent editions have been published in paperback by Penguin in Britain, Sage in the United States. It has proved to be one of the most durable books in organizational studies, with sales over the years of more than a quarter of a million copies. It has been translated into Japanese, Russian, Hungarian, Bulgarian, Romanian and Slovakian. I am told that there is also a pirated version in Chinese, but I have not seen it. Bob dropped out as an author after the third edition, owing to pressure of other work. David and I produced the fourth edition in 1989 after I had spent a happy Christmas vacation unsexing the language: "professional father figures" becoming "professional forebears" etc. We are currently working on the fifth edition for publication in 1996. The book ensures that we keep up with the literature, and involves the interesting decision, which we make after a small survey of our colleagues, of which writers to add and who to drop for each new edition. I used to feel a bit guilty about those dropped, but now we have produced a hardback omnibus edition, *Great Writers on Organizations*, which includes all those writers who have ever appeared.

I do feel that *Writers* is a real contribution to management education in Britain. I continually meet managers who tell me that reading it had started them thinking seriously about organizational issues. Communication with managers has always been an important value for me. I am proud of how many professional and popular articles we have written about our researches, in addition to the academic ones. And I have taken every opportunity to work with practicing managers in research, teaching and consulting.

It was in the mid-1960s that I was asked by Charles Clark of Penguin Books to be the General Editor of a series of entitled Penguin Modern Management Readings. Readers were very popular then in all subjects as the writing of textbooks could not keep pace with the expansion in higher education. I like this sort of editorial role and was pleased to accept. My task was to design the set of titles in the series and then to ask a leading scholar to edit a volume. Of course, getting a "name" to edit a reader is much easier than getting them to write a book, which is why the series went with a swing and my network grew rapidly. So, for example, Igor Ansoff did a reader on business strategy, Vic Vroom and Ed Deci did one on motivation, Warren Bennis and John Thomas on change, Andrew Ehrenberg on consumer behavior, Fred Emery on systems thinking, Dalton McFarland on personnel management, and so on. My own contribution on organization theory is still going strong, and I shall be bringing out the fourth edition in 1997 in parallel with the new edition of *Writers*.

THE FINAL PHASE OF THE I.A. RESEARCH UNIT

The final phase of the I.A. Research Unit took place from 1968 onwards with its splitting up. I am often asked why this happened, with the implication that a successful work group should go on forever. But it is not like that. I am inclined to think that research groups have a life cycle in which they are productive, and the Aston unit was longer lived than most. But people's horizons, opportunities and aspirations change even when they are doing good work, and these form the pull factors. Even in "success" there are many stresses and frustrations which can provide the pushes.

In my case there were two pull factors. First, the attraction of going to a recently established business school in London that was an independent institution wholly devoted to developing research and teaching in the subject. I have a taste for educational innovation and saw that this was a big opportunity. Secondly, I was head-hunted; only the second occasion in my career (so far?) that this has happened to me. Head hunting was unusual in the British academic world, and I took it as a welcome example of the degree of independence which the London Business School (LBS) had obtained from the federal University of London, compared with the workings of the unitary University of Aston (see next paragraph).

The push factor came from the fact that I had unsuccessfully applied for a chair at the University of Aston. It was not that I did not get the job (nobody was appointed), but that the University had adopted a policy which meant that, in principle, I *could not be* appointed. They were advised by an external assessor (in British universities, chair appointing committees always contain professors from other universities) to look to appoint an industrial sociologist who would both build up the organizational work in the management school, *and* develop a sociology department in the new Faculty of Behavioural Science. Apart from the fact that I did not qualify, I fundamentally disagreed with this policy. My view was (and is) that business schools are adequately developed only by academics who are willing to commit themselves in career terms to the new venture. They do not have to cease being sociologists or economists, but their organizational identification must be fully as sociologists in business schools. There was also a second problem: Was it possible to find a leading industrial sociologist who would be prepared to join the management school at Aston? In my view, it was not.

As someone who has spent a large chunk of his life studying the workings of bureaucracies, I have always believed in putting my knowledge to use. I predicted that, having accepted what sounded to those who did not know the field like a viable policy in regard to the chair, it would take the powers-that-be of the University of Aston five years to discover that they could not carry it out. They would then accept the sensible policy of an organizational behavior chair in the management school, and a separate chair to head the sociology department. I left Aston, with considerable regret, because I was sufficiently ambitious not to be prepared to wait those five years. This was one of my predictions which was supported by the data. Precisely five years later John Child was appointed to the chair of organizational behavior in the management school, and in the following year a professor was appointed to head up the sociology department. I acted as an external assessor for both posts.

So, in 1968, I went with John Child and Will McQuillan to the London Business School, Roy Payne following a year later. For quite unrelated reasons, David Hickson and Bob Hinings accepted a two year secondment at the University of Alberta, Edmonton. There David led a group in studies of power, and on his return to Britain, inaugurated the Bradford decision-making program. Bob returned from Canada to the Institute of Local Government Studies at Birmingham, later going back to Alberta to continue his studies of strategic change. At the time, all these departures led to a winding down of the program at Aston, and the fourth generation of Aston researchers operated from other institutions, primarily the London Business School and the University of Alberta. The unit at Aston continued as an administrative entity until 1973, when with a general reorganization of the department the title lapsed. Research at Aston continued in other directions.

THE ASTON RESEARCH PROGRAM

The results of the program have been described in the various books and papers, by us and about us, and need not be repeated here. They have made an impact on the field as is shown by the fact that no less than three papers have been designated "citation classics" by the Institute of Scientific Information on the basis of citation counts. I was awarded a University of Aston DSc (a higher doctorate in the British system) on the basis of my contribution, with Paul Lawrence on the examining board. I regard it as a considerable academic accolade that leading scholars, such as Howard Aldrich, Jerald Hage, Marc Maurice, Bernard Reimann and Bill Starbuck, are prepared to write and publish detailed critiques of the work. I don't agree with all they say, but I don't mind being criticized—I find it more difficult to be ignored.

But the studies have not been ignored—far from it. John Freeman's 1986 "editorial essay" on assuming the editorship of the *Administrative Science Quarterly*, actually mentioned the Hawthorne studies, the American Soldier studies and the Aston studies in the same breath—and managed to take my breath away! Interest in the work stems primarily from the fact that the concepts studied are important to the field. But, in my view, there are two other contributory factors. One is the group basis of the research, which allowed extensions and replications to be undertaken by original members of the group and their "fourth generation" collaborators. The second is the publication of extremely explicit descriptions of the methods of the research. This makes it easier for others to utilize the instruments developed. Many studies around the world have been carried out based entirely on the published methodology. I am most appreciative of the two original editors of the *Administrative Science Quarterly*, Tom Lodahl and Bill Starbuck, who agreed to publish such a degree of detail. John Child and Patricia Clark made a key contribution when they prepared the various data sets using the Aston methodology for deposit in the Economic and Social Research Council (ESRC) data archive for the social sciences at the University of Essex. I have also archived some of the original interview schedules of the Aston and LBS studies in the Open University.

I want to react to two comments on our type of work which never cease to irritate me. One is the suggestion that because we take a functionalist approach, we then use only quantitative data. This is completely wrong, and can be dangerously misleading to the unwary. Some people seem to think that the Aston variables, scales and items jumped straight into our laps, and all we had to do was to start counting! That was not the case; they were the result of a considerable degree of qualitative investigation to understand the relationships involved. How often have I heard Ph.D. students say "I'm going to send out a questionnaire, and then do a detailed study on a small number of cases." This is the wrong way round. Good quantitative work is necessarily based on good qualitative work. As an unreconstructed positivist, I regularly carry out qualitative study, moving on to quantitative designs only if a clear framework of relationships between variables is established which can convincingly be encapsulated in numerical analysis. Oh, and by the way, the Aston structural level studies were based on structured interviews, *not* on questionnaires. These were used only for the group-level studies.

My second irritation is the way the term "radical" has been appropriated by phenomenologists, neo-Marxists, post-modernists, etc., who characterize a positivist, functionalist approach as anti-change. I think that is just not so. If I wanted to change an organization, I would go to someone who understands its structure and functioning, and the levers of change. I would not go to someone who considers that an organization is a domain of discourse or a class-based conspiracy, because it would not give much help in actually changing. So "critical?" Yes, criticism from all is necessary. But "radical?" No way.

THE LONDON BUSINESS SCHOOL

I came to the London Business School when it had been in existence for a couple of years and was still in temporary accommodation. My first post was as Reader and Director of Research of the School, but after a year, on 1st January 1970 to herald the new decade, I was appointed Professor of Organizational Behavior (OB). Although there had been visiting American professors at LBS with this title, including Vic Vroom and Dean Berry (who recruited me) I was the first native Brit in the country to be appointed to a chair in this subject. Tom Lupton had been appointed to the Manchester Business School before me, of course, but his first title was Industrial Sociology. He changed to Organizational Behavior only after I had been appointed. I had been writing quite a lot about the need to forge an interdisciplinary approach to behavioral aspects of management under the title of "organizational behavior," and so was very pleased to have been the first appointment.

My work changed considerably in London, since I had to take on a full teaching and administrative load in addition to research. Natalie too, took up full time teaching with a lectureship (then a senior lectureship) in industrial sociology in the Business Studies Department of the Polytechnic of North London. She was very important in helping me to understand and keep in touch with the polytechnic sector, which then had a large majority of the students in business studies, when, in 1970, I was elected as Chairman of the Association of Teachers of Management. In my three year term as Chairman I had to undertake a more public role: presiding over conferences and committees, developing workshops and training courses for management teachers on professional updating and research, leading delegations to the House of Commons to lobby for the development of management education as a whole, writing to the Education Secretary (Mrs Thatcher) questioning Government policy on "regional management centers," and so on. I also had to learn to make after dinner speeches, which is more difficult than it seems—though I always found the story about how Androcles persuaded the Lion not to have him for dinner got me off to a good start. And my experience of management meant that I was recruited to undertake other non-academic jobs as chairman or consultant in a number of voluntary organizations.

An early job that I had to do at LBS played an important role in my professional thinking. As part of the first review of the master's program, I worked with William Egan in evaluating the validity of the selection test used for the intake. Along with all other leading schools we used the Admissions Test for Graduate Schools of Business of the Educational Testing Service at Princeton (ETS). Since we had only been going three years, only the first year graduates were in jobs. We therefore used the course marks as the criterion measures. We found very high predictive validities from the test scores to the first year marks—in one of the years an overall correlation of over 0.6, and this was before any correction for restriction of range. This was very gratifying to the testers, and our results were used for some years by ETS in their advertising literature. But it worried me. What was the intervening educational process which allowed a selection psychologist to predict so well how somebody would perform more than two years later? How would it be characterized by an educational psychologist? It seemed to me that it had to be a very straightforward process—a "sponge theory" in which education was the one-way transfer of knowledge from the professor to the students, who soaked it up, and then spewed it out in the examinations. I decided that I would throw in my lot with improving the educational process, making it more flexible, more exploratory, more of a two-way exercise, even though this inevitably meant that the predictive validity would fall. I decided that I would rather be an educator than a selector, and since then I have never used a test. I am glad to say that when we reorganized the program, the validities duly went down.

Our Organizational Behavior section at LBS had some interesting teachers. When I first knew Denis Pym he was a mainstream Birkbeck College occupational psychologist. Then he went back home to Australia for a few years. When he returned to London and joined LBS, he had become the most free-wheeling iconoclast that I knew. He seemed to attack everything: organizations, leadership, professionalism, "the domination of the eye over the ear." He was a forceful and effective lecturer and his students liked him. In as far as I could understand him, I regarded his approach as a form of anarchism—which is very liberating, I suppose, for budding executives presumably preparing themselves to oppose anarchy in order to manage successfully.

Andrew Pettigrew was already making a name for himself as a detailed and careful researcher into organizational power and politics, and beginning his magisterial studies of change at ICI, the British industrial conglomerate. He went, in due course, to a chair at Warwick Business School and has built up the Centre for Corporate Strategy and Change there as a leading research group. Charles Handy was the most effective lecturer that we had. Clear, forceful, relevant and inspirational. It is no surprise to me that he has gone on to become a leading British management *guru*—the British version of Peter Drucker. He regularly broadcasts and his books are very popular in challenging received wisdom with visions of the future. Stuart Timperley is still at LBS, where his street-wise organizational wisdom is much appreciated by the experienced executives. And there can't be many academics who have achieved Stuart's distinction of becoming the chairman of a professional football club, namely, Watford Town.

We were later joined by Tommy Wilson (A.T.M. Wilson), one of the most experienced professionals in the field of industrial social science. He had been the first Chairman of the Tavistock Institute of Human Relations, but came to us after having spent a long period as Adviser to the Board on Social Sciences at Unilever. This was probably the most prestigious social science job in British industry, involving continual treading of the corridors of power. He was very shrewd in his understanding of the political processes of organizations, and very powerful in his analysis of policy options. But, maybe because of his long experience as a consultant, I found it very difficult to mobilize his support. For a time we were the only two OB professors in the School. I was the head of department and needed to fight my corner. Tommy always produced clear analyses of the issues, but then usually abstained rather than voted as I felt was required.

At LBS we benefited from a steady stream of outstanding visiting scholars of such enormously varied interests and skills as Chris Argyris, Bob Dubin, Martin Evans, Dale Henning, David Kolb, Ed Lawler, Charles Perrow and Bill Starbuck. I always found it fascinating to watch them in action and to come to terms with their thinking. I like to think I took something from all of them.

I feel particularly proud of two innovations in teaching which I introduced at LBS. Will McQuillan and I developed an Interpersonal Management Skills course as an option in the Master's degree. It became so popular, many students rating it as the best course they had taken in the program, that it was made compulsory, despite my protestations that the act of choice was an essential part of the commitment which made it work. Later Jill Jones and I collaborated, and, after my time there, LBS hired John Harter as a specialist trainer just to carry out this training in a specially designed laboratory.

The second innovation was a course for the OB Doctoral program on the skills of doing research. This course developed from my gradual realization that the education normally given to doctoral students flies in the face of all we know about adult human learning. Typically in Britain, we give doctoral students lectures, tell them to read a lot, and then send them out into the field to "make a contribution to knowledge." Some of them make it, of course, but on the whole it is a recipe for disaster. Effective learning takes place under controlled conditions in which the learner gets plenty of early feedback. I developed a series of graded research exercises, including designing a research study on a specified topic, constructing a questionnaire and testing it out, replicating a published paper, etc. to allow the skills of the research craft to be practiced *before* candidates started on their doctoral projects. Of the series of doctoral students whom I supervised, I still keep in close touch with three: Peter E. Smith who runs the MBA program at the University of Bristol, Moshe Banai of Baruch College, New York, and Cyril Levicki, a management consultant in Oxford.

During this time my international professional network expanded considerably. Rex Adams of Ashorne Hill College, who succeeded me as Chairman of the ATM, asked me to join with him in bidding for a contract to design and run a six-month course in Italy for Italian managers who wished to prepare themselves to become management teachers. We got the contract and this led, over the years, to a considerable amount of management development work in Italy. In due course I was elected a Fellow of the Italian Academy of Business Management. Teddy Weinshall of Tel Aviv University came regularly to LBS and became a friend—particularly after his daughter (Yael-Anna Raveh) did her Ph.D. under my supervision. Teddy helped me to develop my contacts in Israel, and I regularly gave seminars to managers there, many of them jointly with him. I have also been a visiting professor at the Haifa Technion and the Tel Aviv Business School, where Yoram Zeira has been a constant contact. S.R. Ganesh was a doctoral student of mine, and after his return to India he arranged for me to undertake a lecture tour there. I had the honor of being the keynote speaker at the first all-India conference of organizational behavior teachers in Hyderabad in 1979. Other tours there have followed. I also paid a regular series of visits to Hong Kong working with Gordon Redding, giving seminars to managers and collaborating in a joint research project. Regular contact as a teacher and a consultant with practicing managers in both the UK and abroad is important to me as stimulant to my thinking and as a challenge to my teaching.

I had become a member of the Scientific Advisory Committee of the European Institute for Advanced Studies in Management based in Brussels. This enabled me to make contact with many colleagues on the continent. I contributed to a conference, organized by Geert Hofstede and Sami Kassem, on European Contributions to Organization Theory and met a range of new colleagues from different countries there. All of these experiences were sensitizing me to the need to understand cultural differences in organizational functioning. Until then it was David Hickson who had taken the lead in the consideration of cultural differences in the Aston work and its derivatives. The problem, I found, was that "culture" as a explanatory term was a residual category: a rag bag into which it was too easy to throw things which would otherwise be left unexplained. Geert Hofstede's contribution in establishing a framework of four dimensions of societal culture was a major step forward in analyzing the differences. And I find it significant that two of the dimensions map on to Aston dimensions. I have become more and more interested in cultural differences in management, as I feel that we can go beyond description and take a more analytical approach to them.

THE ORGANIZATIONAL BEHAVIOR RESEARCH GROUP

From my arrival at LBS in 1968, Dean Berry and I began building up the Organizational Behavior Research Group with support from the School. After he left to fulfill his nominal potential by becoming Dean Dean Berry of INSEAD (the multi-lingual business school in Fontainbleau, France), the Group obtained a Social Science Research Council program grant entitled

"Organizational Behavior in its Context." This had a number of components. It began with John Child and Will McQuillan working on the "national" Aston project. Roy Payne joined to develop group-level studies on the performance of managerial work groups. Malcolm Warner and Lex Donaldson became "fourth generation" Aston researchers, working on extensions to the program to trades unions and occupational interest associations, in collaboration with Ray Loveridge. Roger Mansfield came to work primarily on managerial careers, but also become a member of the fourth generation making important conceptual contributions both at the structure and group levels. Alan Dale (now of Brunel University) joined to study organizational change and development and Leslie Metcalf (now of the European Institute in Maastricht) to research QUANGOS (quasi non-governmental organizations, such as the economic development committees then set up for particular industries). Brenda Macmillan worked on managerial mobility. Kay Schraer was an excellent secretary. Our size meant that, as is traditional for me anyway, we were housed in a nearby slum—the fine new premises of the business school in Regent's Park being already too small to accommodate researchers.

Good research projects were carried out, as reported, for example, in the second and third volumes of the Aston books and elsewhere. But we were never able to get anything like the degree of group cohesion that we had experienced at Aston. Chris Argyris was a regular visitor to LBS, and on occasion we benefited from his consummate skills as a process consultant when relationships in the research group needed help. After a few years John Child went back to Aston as a professor (i.e., full professor in American terms) and Malcolm Warner went to be professor at the Administrative Staff College, Henley—both meteoric rises: from Senior Research Officer to Professor in one step. Both later went on to develop their many contributions to the field, finally returning to their alma mater, Cambridge University in the Judge Institute of Management Studies; John as the first Guinness Professor of Management Studies. Roger Mansfield, also a LBS Senior Research Officer, paused slightly at Imperial College on the way to becoming Professor and Director of the Cardiff Business School. Ray Loveridge went to a chair at Aston from an LBS lectureship.

In addition to presiding over the whole research program, I was continuing my writing aimed at developing the field of organizational behavior. I had an interesting experience in crossing swords in print with Lyndall F. Urwick. He was the leading British exponent of traditional management theory, and wrote a paper in the then newly established management journal *Omega* attacking social science for confusing what was previously straightforward by misuse of the word "organization." Sam Eilon, the editor of the journal invited me to comment, and I published a short defense of social scientists' contribution to the understanding of management activity. Urwick's reply was a poem called "Lines on D.S.Pugh's Theory of Organization" which begins:

I have no animus
Against the Pughsillanimous,
Nor any intention
To handicap invention.
But D.S. Pugh's semantics
Play such curious antics,
That its hard for a simple mind
Not to be left behind.

It then went on to say that what I had written was nonsense. Since I had gone to some trouble to point out that I did not consider Urwick's views nonsense, only incomplete, I was a bit put out at the time. But still, I suppose it must be some sort of distinction to have a poem about you published in an academic journal.

My specific research interests, in collaboration with Lex Donaldson and Penny Silver, focused on attempting to understand and analyze organizational processes. In the early 1970s we held a conference on "organizational process" to try and get a better understanding of how to analyze the phenomenon. We invited Karl Weick, whose book *The Social Psychology of Organizing* had recently made a big impact. I remember being very impressed that, when I told him that I did not understand the second half of his book, Karl replied that he did not understand it very well either. I thought this must be real innovation, not just a taste for it. The longitudinal study that Lex, Penny and I carried out also reflected my growing interest in organizational change. I was impressed with the ideas that Alan Dale, our most experienced organizational development practitioner, was demonstrating. The excessive emphasis in American OD on interpersonal relationships, always seemed to me to be inadequate. Not surprisingly, I was clear that you had to take the authority structure seriously and look for ways of getting structural change.

That Lex Donaldson left for Australia was a failure on my part. Not for Lex, of course, he has gone on to develop a powerful world class career as an organization theorist based in Sydney. But I could not persuade the Principal of the London Business School that someone would actually leave the School while there was still some chance that he would get an extension to his contract in London. He was therefore not prepared to advance in time the decision on Lex's extension.

I felt that this was typical of LBS at this time. It was at the top of the tree: the leading business school in Britain, the only non-American school consistently listed in the world top ten. It was in a beautiful location in a park in central London, and was a very comfortable place to be. Our Principal was the economist Jim Ball (later Sir James Ball); an excellent leader who had done a magnificent job in revitalizing the School in the 1970s. But by the 1980s, he had stayed too long. (Pugh's rule-of-thumb for successful chief executives has a Macawber-like ring. Aim to stay for ten years. Go after nine: result, sighs of nostalgia. Go after eleven: result, sighs of relief!) I felt that the School was in a rut and, after fourteen years, I was in a rut too. So when the opportunity came in 1983, I was able to exercise a taste for innovation.

OPEN UNIVERSITY, TECHNOLOGY FACULTY, SYSTEMS DISCIPLINE

The United Kingdom Open University (OU) is the leading distance learning institution in the world. It led the way in developing the most important innovation in higher education in modern times: the facility for students working part-time and at home to conduct rigorous university level studies to the same degree standards as at established universities. For the central academic staff it requires a completely different way of working. Teaching does not mean taking a class at nine o'clock on Tuesday morning—there are no students on the campus to teach (except Ph.D. students, of whom more anon). Teaching means participating in a course team which writes course units (as the specially designed workbooks are called), makes audio-visual presentations in collaboration with the BBC (originally as radio and television programs but more commonly now as audio and video tapes), designs the short residential schools, and sets up and participates in computer conferences. In addition to central staff in the "course writing factory" at Milton Keynes, there are also regional academic staff across the country who tutor the students through the courses prepared at the center. It is a completely different way of working from my previous teaching experiences. I was ready for a change and entered into all these activities with a swing.

The reaction in the profession to my move from LBS to the OU, was one of bewilderment—if anything more incredulous than the reaction to my move twenty-five years earlier from Edinburgh University to Birmingham Tech. I was the first—and still, I believe, the only—full professor of the London Business School to move from there to another university. I was moving from the top status institution in my field to a new-fangled set-up which was viewed with much suspicion. For many university teachers, it appears demeaning that the OU requires no previous qualifications at all for entry to its foundation courses. It aims to take people who left school thirty years ago with no qualifications, but who have the ability and the commitment, to obtain a degree. The view is that it is not the input level but the educational process and the output standards which matter. For several years people would come up to me at conferences and say how surprised they were at my leaving LBS. I think many were bemused when I said it was because I have a taste for educational innovation.

I was also changing in another way, too. I was widening my academic interests by joining the Faculty of Technology and accepting the chair of Systems and headship of the Systems discipline. This brought me into contact with a subject that I had previously only been aware of in a general way, but now had to tackle seriously. I should say that the University had defined the post in the widest possible terms, and, indeed, the advertisement for the job had specifically stated that experience in organizational development was an appropriate basis for applying. I underlined at my interview that there was no point in appointing me unless the OU was intending to develop management studies in a major way, and the Vice-Chancellor confirmed that this was so.

My Inaugural Lecture linked both interests, being on "The Management of Complex Systems." In British universities a new professor on appointment gives a lecture which is open to the whole university and, indeed, to the public. But I had given public lectures before and did not come to the OU to repeat myself. My Inaugural Lecture was a fifty minute television program; a first even for the OU. It is a lecture but, since I had the resources of the BBC behind me, I was able to illustrate it with footage from programs as disparate as Henry Ford's Model T assembly line, the "Yes, Minister" sitcom, and specially filmed interviews about their research with Frank Heller and David Hickson. I was talking direct to camera for less than half of the time. The program was shown several times on television, and after some years was even revived as a "golden oldie."

The fields of application of the systems discipline were very wide ranging. There was research on bio-systems, catastrophic systems failures, manufacturing systems, energy systems, and the functioning of worker co-operatives. The group, using the ideas of Sir Geoffrey Vickers, Stafford Beer, Peter Checkland, took what I would regard as a cybernetic engineering approach to systems. It always surprised me that my colleagues did not draw on the biological systems thinking of von Bertalanffy, and Emery and Trist. Even the bio-systems unit seemed to me to be concerned with an engineering approach to biological change. And, with few exceptions, the discipline's interest in management was confined to systems design. As head of department my involvement in these activities was limited to encouragement and criticism. My own writing at this time was focused on reviews and re-evaluations of past work, such as that for the fascinating conference on *Beyond Method* which Gareth Morgan organized at York University, Toronto.

The only students on campus are those studying for Ph.D.s: at this level the OU becomes like an ordinary university with personal supervision. I found, as is so often the case in Britain, that these students were not well inducted into the nature of a Ph.D. They did not know how the process works, and what was expected of them. Their supervisors seemed to think that "they will pick this up as they go along" and left it at that. At the London Business School, in addition to my skills course for the OB doctoral students, I had inaugurated during my term of office as Chairman of the Doctoral program, a course for all students on "The processes of Ph.D.-getting." It dealt with such usually neglected topics as the meaning of a doctorate, the form of the Ph.D. thesis, and perils to avoid. When I came to the OU, I began to offer this course to the Technology Faculty and then to the whole university.

While I was still at LBS, I was asked to be the external examiner for a Ph.D. thesis on concept development in doctoral students by Estelle Phillips. When I came to the OU, I found that Estelle had carried out an evaluation study of the University's doctoral supervision system. Clearly we had complementary interests in this process and I suggested that we collaborate on a book which would help students to understand and manage their way successfully through their doctoral studies. She agreed, and in due course Phillips and Pugh: *How to Get a Ph.D.* came into being. Now in its second edition, it has been a very successful book both in Britain and abroad. It is distinctive in that it concentrates on the processes involved, and being applicable to all subjects is as avidly read by scientists and engineers as by economists and historians. "How *not* to get a Ph.D.—seven tried and tested ways" and "How to manage your supervisor" are popular chapters, as is the one on "How to supervise" which is addressed to the supervisor, the other key partner in the enterprise. Since it appeared, Estelle and I have been running a sort of unofficial counselling service. Students, supervisors, even Deans, call us up for advice on difficult problems. Some of the ways in which research students are treated are truly hair-raising, and I often think of us as the pathologists of the doctoral process. The second edition of the book contains a detailed chapter on "Institutional Responsibilities" which we hope will contribute to improving the standards of provision.

I was also much involved at this time in the battles to establish a business school in the Open University on what I would consider to be a proper basis. The teaching of management had begun in the Continuing Education program of the University under the direction of Brian Lund. I acted as a tutor in the London Region for the introductory course entitled "The Effective Manager" for which Brian had been responsible, with contributions from, among others, Charles Handy. This excellent course was very popular, and further courses were introduced. But there were inevitable disagreements about the nature of the qualifications, the need for an MBA degree, the necessity to establish a school as a full faculty in the University able to appoint professors, and so on. These were duly hammered out, and in 1988 the business school was established as the Faculty of Management of the Open University.

Having established a school, the University's first task was to appoint a Dean. With many misgivings on Natalie's part, she is always much more realistic than I am, I applied for the post. In what I now regard as a providential escape, I was not appointed. Looking back I realize that I was too influenced by the independent situation of the London Business School: I conceived of the

Deanship as something like the Principal's role there. In fact, for most deans of British university business schools, the main job appears to be fighting your own university—and I would have got very frustrated with that.

The OU appointed Andrew Thomson of the University of Glasgow as Dean. Andrew and I were already working together on an Economic and Social Research Council working party for the establishment of large scale databases in management studies. He immediately asked me to join him in the business school. I was lobbied by Geoff Peters, the Dean of the Technology Faculty, to stay there. But there was no doubt where my heart lay—and anyway I have a taste for new expanding enterprises, rather than mature ones.

THE OPEN UNIVERSITY BUSINESS SCHOOL

So here we are again in temporary accommodation in Stony Stratford while our campus building in Milton Keynes is being erected. The Open University Business School (OUBS) was not in a slum this time but in a suite of offices in a faceless modern block. In 1988 it was a hive of activity: new structures, new subjects, new courses, exponentially expanding numbers of students. In four years it became, in terms of student numbers, the largest business school in Europe, probably in the world. That's the nature of successful distance learning. And then we began expanding into Western Europe, followed, with the help of the "know-how fund" by expansion into ex-communist Eastern Europe, and then into the Pacific Rim. The whole exercise had an attractive 1960s feel about it, and we were able to do it in the late 1980s because the developments were funded from fee income with only minor government financial support.

In this welter of teaching activity, I was appointed as Director of Research with a brief to introduce elements of a research climate, establish research activity, and inaugurate a doctoral program. Some financial resources were available, but as anyone who has been in this situation knows, it is not money that is the academics' scarce resource, but time. Over the years we have established a number of University recognized research groups built around committed researchers: they include voluntary sector management, small business management, strategic management, information management, human resource management and, my particular concerns, international management and management history.

I also inaugurated the OUBS doctoral program, but it has, as yet, a small number of students because we cannot go beyond the capacity of our faculty to supervise, and their research experience is still being steadily built up. Then the ESRC established a national Management Teaching Fellows program to give encouragement to beginning academics to enter the field of management studies. The program gave opportunities for the participants to have a reduced teaching load, while they obtained training in teaching and research. During the years of the scheme's operation, the OUBS obtained over a dozen such fellowships, one of the largest allocations of any business school. I was responsible for designing and managing the program of activities for them; in later years in collaboration with Jacky Holloway, one of the first graduates of the scheme.

While Andrew Thomson was still at Glasgow, he had agreed to edit the *Newsletter* of the newly formed British Academy of Management (BAM). Andrew edited the first issue at the OUBS, but then, because of pressure of work, he asked me to take it over. With my previous experience of the ATM *Newsletter*, I can't say that I didn't know what I was letting myself in for. But I enjoy this sort of editing. It is proactive. It requires the editor to go out to colleagues, asking, and then nagging, them to write informative or provocative articles. I was able to make full use of my network to establish a regular publication which BAM members told me they positively looked forward to reading. I established a number of regular features: "BAM Soapbox" to stir things up by "banging the drum at the top of your voice about a bee in your bonnet;" "BAM Impact" for introducing your colleagues to an important book in your field; "BAM Focus" for publicizing the research being carried out in your school. Perhaps the most popular feature was the gossip column written by "Stony Stratford." This managed to comment on the passing scene to such good effect that even some of my American colleagues told me they read it to discover what was happening in British academic management. After three years and ten issues, I felt that I had done my stint and handed the *Newsletter* on.

THE INTERNATIONAL MANAGEMENT RESEARCH GROUP

By the time I moved into the Open University Business School in 1988, cross-cultural comparisons in management had become a main focus of my research interests and I decided to take the title of Professor of International Management—the third chair that I have occupied. David Hickson had put us in touch with Enzo Perrone of Bocconi University, Milan—the leading business school in Italy. They had been conducting comparative longitudinal studies of the structure and functioning of Italian organizations. They were setting up a European network to carry out cross-cultural studies and were seeking a British participant for the group. I agreed to join, and set up the OUBS International Management Research Group which became the British member of the "International Organization Observatory." The other members of the IOO are from France (Gilles van Wijk, Paris) Spain (Josep Baruel, Barcelona) the Netherlands (Arndt Sorge and Mariëlle Heijltjes, Maastricht) and Germany (Christian Scholz, Saarbrucken). We are setting up a European data base that will allow continent-wide comparisons. The first round of data have been collected, the second is in train. Geoff Mallory, Timothy Clark and I work on this project. It requires considerable investment in getting research collaboration across many countries, but we consider that this is the only way to get good cross-cultural comparative data. I feel it appropriate that early papers from this work have appeared in German, and in the *Festschrift* for my friend and neighbor Frank Heller of the Tavistock Institute, who has had such a long experience of cross-cultural research through the Industrial Democracy in Europe program. A further result of the IOO collaboration is a book on "European Perspectives on Human Resource Management," edited by Timothy Clark, which explores the interestingly different ways in which this subject is conceptualized and practiced in the various European countries.

With my colleague Dagmar Ebster-Grosz, I have also been conducting a more specific project on Anglo-German business collaboration. This was designed as joint venture with the University of the Saarland and involved interviews with the chief executives of German subsidiaries in the UK and British subsidiaries in Germany. The two research teams made their data available to each other, but, because of a major divergence of opinion, made their analyses separately. In our view, the German analysis is overly quantitative in a way which is unjustified by the nature of the data. The British results are based on content analysis of the interviews and illustrative quotations. This harks back to a point I made earlier about the need for a strong qualitative understanding before undertaking quantitative analysis. Our results will be published in a book in 1996. We are currently preparing to carry out an analogous project on Anglo-French business ventures.

My work in this field has linked up with David Hickson's. He proposed that we collaborate on a book which would review the impact of societal cultures on management activity in different countries around the world. The distinctive concept was that it would be organized by country, so that a reader could look up a particular country and obtain a summary of what is known of its management approach. Since collaborating with David is one of life's pleasures, I enthusiastically agreed. But I was hopelessly over-optimistic on time scales, and David had to do a lot of chasing up. But still, the book, *Management Worldwide*, was published by Penguin in 1995 and appears to be selling well. We are hoping that it, too, will run to further editions, like our previous Penguin book.

MANAGEMENT HISTORY RESEARCH GROUP

A major recent interest of mine, as befits a geriatric professor, is in management history. This was stimulated when I met a phenomenon: E.F.L. Brech. Edward Brech has been writing about management for the last fifty years. He collaborated with L.F. Urwick on the classic British three volume work which appeared in the 1940s on *The Making of Scientific Management*. He wrote *The Principles and Practice of Management* and other books, and is in the Pugh and Hickson *Great Writers on Organizations* omnibus. And all this while working as a manager and a management consultant; he has never held an academic post. In the years since his retirement he has been writing a history of the development of management and the management professional institutes in Britain. He was looking for an academic link and, strangely, had found it difficult to find one.

I consulted with Andrew Thomson and we agreed to set up a Management History Research Group in the Business School to develop this work. When Andrew finished his stint as Dean, he became the head of the group. Edward became a Visiting Research Fellow, and, under my supervision, developed his work on "The concept and gestation of a professional institute of management in Britain, 1902—1949" into a thesis. In 1994, at the age of 85, he was awarded a Ph.D. for his study. Edward continues to work in this field, and, under Andrew's leadership, the group's work continues to expand.

My own interests are in the history and development of management ideas. My contribution, so far, has been as the series editor for Dartmouth Publishing's *The History of Management Thought*. Many new universities cannot make available a historical framework for management studies because they do not have back runs of important journals, and are not in a position to obtain them. A set of readers, in which a leading scholar chooses key articles in the field to demonstrate its historical development, seemed a useful contribution to make. This is what the series sets out to do. So we have John Miner's choice on Administrative and Management Theory, Lyman Porter's and Greg Bigley's on Human Relations, Sam Eilon's on Management Science, Lex Donaldson's on Contingency Theory, etc. The series ranges from Early Management Thought (Dan Wren) to Post-modern Management Theory (Marta Callas and Linda Smircich). It is going well and we intend to expand it. I shall be doing a volume on cross-cultural comparative management.

ENVOI

We have now come to the present and near to the end of the month when I retire. It tickles my taste for innovation that I shall be doing something on my last working day that I have not done before—acting as an external examiner for a continental Ph.D. This is not like a small British affair, with the two examiners, the supervisor and the candidate huddled together in an office. I am one of a committee of eight professors sitting in full academic dress in the main lecture hall of the University of Limburg, who ask questions of the candidate, Mariëlle Heijltjes. She has to defend her thesis in public for one hour, while her colleagues and friends look on. I don't know about the candidate, but I am looking forward to it.

I am also looking forward to the *Festschrift* called *Advancement in Organizational Behaviour* which Timothy Clark and Geoff Mallory are editing on the occasion of my retirement, and I understand that many friends have been kind enough to contribute to it. And I am delighted that Rosemary Thomson and members of the course team of "The Effective Manager" course have established the annual Derek Pugh Prize, to be awarded to the best student among the two thousand or so managers who take this course every year.

With the encouragement of the present Business School Dean, David Asch (a doctoral student of mine), I am continuing my involvement from next month on a part-time basis with the international management and management history research groups at the OUBS. Then there are the new editions of the books to write, and new ideas for book series to develop. I know it sounds suspiciously like workaholism, but I have to say that I have enjoyed myself at work—even in the battles, and I have had my fair share of those. So there is a natural attraction to carry on. And anyway, my view has always been that while we may have made some progress with the various research programs, we are still trying and "the best is yet to be." And right on cue, the British Academy of Management's 1996 Annual conference is being held at Aston with the theme "30 years on... What have we learned?" The first generation Aston researchers have been asked to be keynote speakers.

But, as I said at the beginning, things will change. Natalie has already taken early retirement, and we have a range of non-work activities—familial, social, cultural, educational—that have been put on relative hold in the recent years. These can now be better nourished. I am, after all, still likely to indulge my taste for innovation by doing something a little different.

PUBLICATIONS

1954

A note on the Vorhaus Rorschach-Configurations of Reading Disability. *Journal of Projective Techniques, 18*, 478-480.

1957

With R.M. KcKenzie. Human aspects of inspection in industry. *Journal of Institution of Production Engineering, 36*, 378-387.

1959

With C. Gordon & A.R. Emerson. The age distribution of an industrial group (Scottish Railwaymen). *Population Studies, 12*, 223-239.

With C. Gordon & A.R. Emerson. Patterns of sickness absence in a railway population. *British Journal of Industrial Medicine, 16*, 230-243.

With C. Gordon & K. Levy. Sickness absence among railway clerical staff. *British Journal of Industrial Medicine, 16*, 269-273.

With J.M. Fraser. Wastage in a workshop supervision course. *Technology.*

1960

From business game to management exercise. *British Association for Commercial and Industrial Eduation Journal, 14*, 140-143.

1961

Effective staff selection procedures. *Journal of the Institute of Office Managers, 15*, 197-198.

1962

The industrial administration research unit. *Management Thinking, 1*, 10-14. (I.A. Department, Birmingham College of Technology).

Management Studies without Research? *Bulletin of the Association of Teachers of Management, 6*, 15-18.

1963

With D.J. Hickson, C.R. Hinings, K. MacDonald, C. Turner, & T. Lupton. A conceptual scheme for organizational analysis. *Administrative Science Quarterly, 8*, 289-315.

1964

With C.R. Hinings. *Developments in the empirical study of bureaucracy*. Paper presented to the British Association for the Advancement of Science, Annual Meeting, Southampton.

The structure of industrial enterprise in industrial society—A comment. *Sociological Review Monograph, 8*, 63-64.

With D.J. Hickson & C.R. Hinings. *Writers on organizations*. London: Hutchinson.

1965

With D.J. Hickson. The facts about "bureaucracy." *The Manager.*

Games and exercises: A comment on terminology. In E.A. Life & D.S. Pugh (Eds.), *Business exercises: Some developments.* Association of Teachers of Management, Occasional Paper No. 1. Oxford: Basil Blackwell.

The profession of management. *Association of Teachers of Management Bulletin, 16*, 2-4.

T-group training from the point of view of organization theory. In G. Whittaker (Ed.), *Group dynamics for management education.* Association of Teachers of Management, Occasional Paper No. 2. Oxford: Basil Blackwell.

Aims and methods of modern management education. *Scientific Business 3*, 258-266.

With J.N. Fairhead & W.J. Williams. *Exercises in business decisions.* London: English Universities Press.

1966

The social science approach to management. *Scientific Business, 4*, 23-31.

The first European summer school in social psychology: Some impressions. *Bulletin of the British Psychological Society, 19*, 35-37.

Modern organization theory: A psychological and sociological study. *Psychological Bulletin, 66*, 235-251.

The profession of management. In D. Pugh (Ed.), *The academic teaching of management.* Association of Teachers of Management, Occasional Paper No. 4. Oxford: Basil Blackwell.

The teaching of management theory. In D. Pugh (Ed.), *The academic teaching of management.* Association of Teachers of Management, Occasional Paper No. 4, Oxford: Basil Blackwell.

Role activation conflict: A study of industrial inspection. *American Sociological Review, 31*, 836-842.

Organizational problems of inspection. *Journal of Management Studies, 3*, 256-269.

1967

With C.R. Hinings, D.J. Hickson, & C. Turner. An approch to the atudy of bureaucracy. *Sociology, 1*, 61-72.

With R.L. Payne & J.H.K. Inkson. Extending the occupational environment: The measurement of organizations. *Occupational Psychology, 41*, 33-47.

With R.L. Payne, D.J. Hickson, & J.H.K. Inkson. *Social behaviour in organizations.* Paper presented at the Annual Conference of the Social Psychological Section of the British Psychological Society, Oxford.

1968

With D.J. Hickson. The comparative study of organizations. In D. Pym (Ed.), *Industrial society: The social sciences in management*. London: Penguin books. (Reprinted in French in *Synopsis*, Journal of the Belgian Productivity Association.)

With D.J. Hickson. A dimensional analysis of bureaucratic structures. In R. Mayntz (Ed.), *Burokratische Organization.* Berlin: Kiepenheuer and Witsch (in German).

With D.J. Hickson, C.R. Hinings, & C. Turner. Dimensions of organization structure. *Administrative Science Quarterly, 13*, 65-105.

With J.H.K. Inkson & D.J. Hickson. *Administrative reduction of variance in organizational behaviour*. British Psychological Society Conference Paper (reprinted in Pugh & Payne, 1977).

With D.C. Pheysey. Some developments in the study of organizations. *Management International Review, 8*, 97-107.

1969

With P. Levy. Scaling and multivariate analysis in the study of organizational variables. *Sociology, 3*, 193-213.

With D.J.Hickson, C.R. Hinings, & C. Turner. The context of organization structures. *Administrative Science Quarterly, 14*, 47-6l.

With D.J. Hickson & C.R. Hinings. An empirical taxonomy of structures of work organizations. *Administrative Science Quarterly, 14*, 115-126.

Organizational behaviour: An approach from psychology. *Human Relations, 22*, 345-354.

With D.J. Hickson & D.C. Pheysey. Operations technology and organization structure: An empirical reappraisal. *Administration Science Quarterly, 14,* 378-397.

With D.C. Pheysey. A comparative administration model. In A. Neghandi (Ed.), *Comparative administration research.* Kent, OH: Kent State University Bureau of Business Administration.

Organization theory. In T. Kempner (Ed.), *A student's guide to management studies*. Association of Teachers of Management, Occasional Paper No.6. Oxford: Basil Blackwell.

Management education in Britain. In E. Blishen (Ed.), *Encyclopaedia of education*. London: Blond.

1970

With D.C. Pheysey & D.J. Hickson. Organization: Is technology the key? *Personnel Management*, February, 21-26.

With J. Child. How to measure organization. *Management Today*, February, 127 -129.

The organization of the marketing specialisms in their contexts. *British Journal of Marketing, 4*, 98-105.

With J.H.K. Inkson & D.J. Hickson. Organization context and structure: An abbreviated replication. *Administrative Science Quarterly, 15*, 318-329.

1971

[Editor] *Organization theory: Selected readings*. London: Penguin Books.

With R.L. Payne. Organizations as psychological environments. In P. Warr (Ed.), *Psychology at work*. London: Penguin Books.

With D.C. Pheysey & R.L. Payne. Influence of structure at organizational and group levels. *Administrative Science Quarterly, 16*, 61-73.

With D.C. Pheysey & R.L. Payne. Organization structure, organizational climate and group structure: An exploratory study of the relationships in two British manufacturing companies. *Occupational Psychology, 45*, 45-51.

Organizational behaviour in its context. *Social Science Research Council Bulletin, 12*, 8-9.

With D.J. Hickson & C.R. Hinings. *Writers on organizations* (2nd ed.). London: Penguin Books.

With W. Egan. Selection for the MSc programme at the London Graduate Business School. In K.M. Miller (Ed.), *Managers in the making*. London: Independent Assessment and Research Centre.

1972

With D.J. Hickson. Causal inference and the Aston studies. *Administrative Science Quarterly, 17*, 273-276.

Developments in organization theory. In R. Piret (Ed.), *Proceedings of XVIIth International Congress of Applied Psychology* (pp. 833-836). Brussels: Editest.

1973

Measurement of organization structures. *Organizational Dynamics, 1*, 19-34.

Colonel Urwick and organization. *Omega, 1*, 347-352.

1975

With R. Mansfield & M. Warner. *Research in organizational behaviour: A British review*. London: Heinemann.

Organizing for people. *New Behaviour, 22*(May), 205-207.

1976

With D.J. Hickson. *Organizational structure in its context: The Aston programme I.* Aldershot: Gower Publishing.

[Edited] With C. R. Hinings. *Organizational structure extensions and replications: The Aston Programme II.* Aldershot: Gower Publishing.

With R.L. Payne. Organizational structure and climate. In M. Dunnette (Ed.), *Handbook of organizational psychology.* Chicago: Rand McNally.

The "Aston" approach to the study of organizations. In G. Hofstede & M.S. Kassem (Eds.), *European contributions to organization theory.* Amsterdam: Van Gorcum.

Going longitudinal. In C. Brown et al. (Eds.), *The access casebook.* Stockholm: THS.

Motive power. *Accountancy Age, 7*, 6 August.

With L. Donaldson & P. Silver. *A comparative study of processes of organizational decision-making: A preliminary report.* Paper given to Conference on Current Studies on Work Organizations, Berlin, October.

With D.J. Hickson & C.R. Hinings. *Writers on organizations* (2nd ed.). Tokyo: Tuttle-Mori. [in Japanese.]

1977

[Edited] With R.L. Payne. *Organizational behaviour in its context: The Aston programme III.* Aldershot: Gower Publishing.

Communication breakdown. *Accountancy Age, 8*, 14 January.

1978

Understanding and managing organizational change. *London Business School Journal, 3*, 29-34.

1979

Effective co-ordination in organizations. *Society for Adavanced Management Journal*, Winter, 28-35.

1981

The Aston programme: Retrospect and prospect. In A. Van der Ven & W.F. Joyce (Eds.), *Perspectives on organization design and behaviour.* New York: Wiley.

Rejoinder to Starbuck. In A. Van der Ven & W.F. Joyce (Eds.), *Perspectives on organization design and behaviour.* New York: Wiley.

1983

Studying organizational structure and process. In G. Morgan (Ed.), *Beyond method: A study of social research strategies.* Beverley Hills, CA: Sage Publications.

With D.J. Hickson & C.R. Hinings. *Writers on organizations* (3rd ed.). London: Penguin Books.

1984

[Editor] *Organization theory: Selected readings* (2nd ed.). London: Penguin Books.

1985

Management classics. In P. Braithwaite & B. Taylor (Eds.), *The good book guide for business.* London: Penguin Books & New York: Harper and Row.

International perspectives. Unit 16 Open University Course, "Managing in organizations" (T244). Milton Keynes: The Open University.

With D.J. Hickson & C.R. Hinings. *Writers on organizations* [American ed.]. Beverly Hills, CA: Sage.

What is research? In A. Chapman (Ed.), *Management research and management practice.* Association of Teachers of Management, Focus Paper. 3-6.

1986

With G. Redding. The formal and the informal: Japanese and Chinese organization structures. In S. Clegg, D. Dunphy, & G. Redding (Eds.), *The enterprise and management in East Asia.* Hong Kong: Hong Kong University Press.

Modern classics in organization theory. *Management and Labour Studies, 11,* 111-114. [Xavier Labour Relations Institute, Jamshedpur India.]

1987

With E.M. Phillips. *How to get a Ph.D.* Milton Keynes: Open University Press.

Organizational development. Block 4 Open University Course, "Planning and managing change" (P679). Milton Keynes: The Open University.

1988

The Aston research programme. In A. Bryman (Ed.), *Doing research in organizations.* London: Routledge.

1989

Systems and organizations. In R.E. Flood (Ed.), *Systems prospects: The next ten years of systems research.* London: Plenum Publishing.

With D.J. Hickson. *Writers on organizations* (4th ed.). London: Penguin Books & Newbury Park, CA: Sage.

[Edited and Introduction] *The production-inspection relationship* (by R.M. McKenzie, deceased). Edinburgh: Scottish Academic Press.

1990

[Editor] *Organization theory: Selected readings* (3rd ed.). London: Penguin Books.

The convergence of international organizational behaviour? Open University Business School Working Paper No. 2/90. [Reprinted in T. Weinshall (Ed.), *Culture and management.* Berlin: de Gruyter Publishers, 1993.]

1991

Foreword to *Organizational behaviour* by A. Huczynski & D. Buchanan (2nd ed.). London: Prentice-Hall.

1993

With D.J. Hickson. *Great writers on organizations: The Omnibus edition.* Aldershot: Dartmouth Publishing.

Organizational behaviour. In W. Outhwaite & T.B. Bottomore (Eds.), *Dictionary of twentieth century social thought.* Oxford: Basil Blackwell.

1994

With E.M. Phillips. *How to get a Ph.D.* (2nd ed.). Buckingham: Open University Press.

1995

Culture. Block 5 Open University Course, "International enterprise" (B890). Milton Keynes: The Open University.

With C. Mabey. Strategies for managing complex change. Unit 10 Open University Course, "Managing development and change" (B751). Milton Keynes: The Open University.

With D.J. Hickson. *Management worldwide: The impact of societal culture on organizations around the globe.* London: Penguin Books.

International management. In N. Nicholson (Ed.), *Encyclopedic dictionary of organizational behavior*. Oxford: Basil Blackwell.

Organizational design & organizational development. In P. Forrest (Ed.), *Croner's A-Z guide for HRM professionals*. London: Croner Publications.

With T. Clark & G. Mallory. Struktur und strukturelle Änderungen in europäischen Unternehmen des Produzierenden Gerwerbes: Eine vergleichende Studie. In C. Scholz & J. Zentes (Eds.), *Strategisches Euromanagement*. Stuttgart: Schäffer-Poeschel Verlag.

1996

With T. Clark & G. Mallory. Organization structure and structural change in European manufacturing organizations. In P.J.D. Drenth, P.L. Koopman, & B. Wilpert (Eds.), *Organizational decision-making under different economic and political conditions*. Amsterdam: North Holland Press.

With D. Ebster-Grosz. *Anglo-German business collaboration: Pitfalls and potentials*. Basingstoke: Macmillan.

With D.J. Hickson. *Writers on organizations* (5th ed.). London: Penguin Books & Newbury Park, CA: Sage.

1997 (to appear)

[Editor] *Organization theory: Selected readings* (4th ed.). London: Penguin Books.

With D.J. Hickson. Organizational convergence. In M. Warner (Ed.), *International encyclopedia of business and management*. London: Routledge.

Never Say Never!

JOHN W. SLOCUM, JR.

On my office door, there is a collage of golf adages and pictures along with this statement: "If you want something badly enough, you are only half alive without it, no matter what the personal sacrifices are needed to get it." This statement has been instrumental in shaping my personal and professional career strategies for the past 40 years or so, for it indicates not only that tradeoffs and sacrifices are needed to achieve any goal, but also, without goals, life is less than it could be. Early childhood experiences often shape future behaviors and several key events prior to my entering college shaped my behavior.

THE EARLY YEARS

It was a hot and muggy September day in Harlem right before World War II that I was born. My father was a copy editor for the *New York Times* and my mother a music teacher, with her degree from Julliard in piano. As my academic career progressed, his influence on my work habits was profound. For example, he instilled me with the desire to complete work timely and write with a message to communicate. While attention to grammar was important, getting the message on to paper was critical. He also was able to "think" at the typewriter and told me that excellent writers developed this skill. Therefore, very early in my career, I learned to type and have been composing at the key board throughout my career.

Management Laureates, Volume 4, pages 277-311.

ISBN: 1-55938-730-0

As the country prepared for war, we moved from New York City to Bergenfield, New Jersey, a commuting distance from 42nd Street, the headquarters of the *New York Times*. The George Washington bridge was open, but ferry boats between New Jersey and New York City still provided daily transportation for my father. I still can remember the vast openness of New Jersey and seeing the lights of the George Washington bridge as a boy. My sister was born two years later, during the war. Gasoline was in very short supply and walking with my mother and sister to the butcher shop was almost a daily chore. We saved fat from occasional meat and turned that into the butcher shop for credit. Squeezing the "red" dot on the oleo plastic container and listening to "The Shadow" and "Suspense Theater" on a crystal radio provided hours of entertainment. Naturally, we all walked to school—rain, snow, heat and sun.

Given my mother's gift for music, piano lessons were suggested. Not gaining much satisfaction from that experience, I changed to the trumpet and became quite talented. Playing solos in school affairs was satisfying and something that I could do well. To accomplish these tasks, took discipline. This sense of discipline has been an important part of my life. I pride myself in being self-disciplined and never missing important engagements or dates. Working over holidays to reach a deadline that was set for students or clients, not playing golf with friends on Friday afternoons to complete chapters of a book or revise a manuscript, and postponing personal vacations requires discipline and a delaying of personal gratification. Even today, all students' term papers and exams are graded within 72 hours. Reviews for journal editors are routinely returned within 48 hours, unless I'm out of the office. Students works are returned with critical comments, and grammatical corrections.

A turning point in this early development stage was in the 9th grade. A Spanish teacher and I had a heated argument over a grade. When the emotional debate concluded, it was decided that I needed to enroll in another high school. In the fall of 1955, I was granted a scholarship at Admiral Farragut Academy in Pine Beach, New Jersey, approximately 130 miles south of Bergenfield. This small school stressed academics, self-discipline, and team work. Because the school had small (my graduating class had 46 boys) classes, cadets were encouraged to play two varsity sports and engage in a variety of extracurricular activities. This preparatory school didn't permit cadets to go home except for national holidays/vacations, and pointed its graduates toward attending either the Naval Academy or West Point. It was during those three years that I developed a high degree of self-reliance and self-discipline. Unfortunately, I was injured in a wrestling match during January of my senior year and couldn't pass the physical examination for the Naval Academy. I remember my depressed feelings and my father challenging me to "pick up the pieces" and move on. We took a trip to visit several small liberal arts colleges and I chose Westminster College. Why Westminster College? First, it was tucked away in the mountains of Pennsylvania. Because I had lived away from home during high school, the distance between college and home was not an issue. Second, the swimming coach indicated that I could "swim" on the team, which was composed mainly of freshmen. The challenge of making the team was exciting. Third, with small classes, I knew that I wouldn't be lost in the crowd and could make an impression on others.

During my sophomore year, I decided to become a professor because of the mentoring of Sam Sloan, a faculty member who taught economics, among other classes. Being at a small liberal arts college, Sam was assigned to teach a variety of different business courses (for example, production, marketing, management, policy). The reality of Sam's classroom never matched the course description nor its purported content. Sam encouraged his students to develop a philosophy of life that would enable them not only to manage themselves, but to lead others. Sam would meet students at Isley's soda shop and engage them in a discussion of current world and college events. He really nurtured students and tried to develop each student to their full potential.

My grades were only fair because of my active social life and inability to get enthusiastic about "trivial" courses during my freshman year. During my senior year, however, I made the Dean's List. Upon graduation in 1962, I enrolled at Kent State University to pursue an MBA, while working in the Industrial Relations Department of the B.F. Goodrich Company in Akron, Ohio.

GRADUATE SCHOOLS

It was during the first semester at Kent State that I met Don Hellriegel in Arlyn Melcher's organization theory class. What was most remarkable about the class was that any of us survived. Arlyn was still working on his dissertation from the University of Chicago and was ill-prepared to teach MBAs. On the first day, he wrote two blackboards full of assignments. Someone asked him if this was the assignment for the term and he replied that all this reading was due for the next week. He also indicated that we would have to find copies in the library because he didn't have time to put much of this on reserve. Grades would be determined by how well you were able to defend the arguments posed by class members regarding a particular assignment. The presenter couldn't earn a grade for the class, but the class member posing the question would be graded on the insightfulness of the questions. We began with more than 30 students, 24 dropped after the first class. Don and I stayed and quickly became good friends.

We took several more classes together before Don graduated and took a job at Cleveland Electric Illuminating Company. I remained at B.F.Goodrich and in the winter of 1964, decided to leave B.F.G. and applied for the doctoral program at the University of Washington. A former Kent State MBA student, Fred Finch, was pursuing his degree at Washington and promised to get me a fellowship if I wanted to attend. While working on my MBA full-time, my social life was almost exclusively confined to either going to class or the library. On one fateful day , I met Gail Gustin in the Kent State library. In February, 1964, we decided to get married. I told Don that we were leaving for the University of Washington so that I could pursue my doctoral degree. Gail had gotten a teaching position in Seattle paying $4,900 and with my fellowship of $300 per term, we were off to Seattle for a three year hiatus. We figured that if I did not make it through the program, we would live in Seattle and I would get an industrial relations position with Boeing. Sharing our thoughts with Don, he said "I'll never do that." Two years later, he called me and asked that I help him get a fellowship to attend Washington's doctoral program. Today, we laugh about that comment and in the preface to the fifth edition of our *Management* book I wrote, "Never say never."

At the University of Washington, the faculty was superb. Monty Kast, Jim Rosenweig, Bud Saxberg, Barry Knowles, Bill Scott, Wendell French in the business school along with Allen Edwards and Ezra Stotland in social psychology greatly shaped my thinking about teaching and research. Stotland would invite students to eat lunch with him, which meant chewing on some raw cabbage and carrots. Through these informal lunches, he imparted his methods of teaching, which paralleled Sam Sloan's. That is, students learn by questioning conventional wisdom and developing their own idiosyncratic paradigms. Through the development of his own paradigms, he was able to relate esoteric social psychological concepts to everyday experiences that students could comprehend and remember.

As a research assistant to Saxberg, I learned the value of clear writing and to carefully and fully explicate your arguments prior to committing yourself to a perspective. We drafted a manuscript seven times, had colleagues read and comment on it, before sending the manuscript to *Management Science*. This was my first publication.

The Academy of Management meetings were held in conjunction with the Allied Social Sciences meetings in December, 1966 in San Francisco. Gail and I had decided that we wanted to return to the east coast and began considering schools in that part of the country. Dale Henning, a faculty member at the University of Washington, was a former University of Illinois, graduate school classmate of Max Richards, who was at Penn State and also the President of the Academy of Management. It was during a cable-car ride in San Francisco that Dale introduced me to Max . We had a brief interview on the cable car. The next day Max called and said that if I was in the State College, PA area, call him for an interview. No one is ever in the State College area unless they plan to be there!!!

It was an interesting time for Gail and I because my dissertation, "Group Cohesiveness: A Salient Factor Affecting Students' Academic Achievements" was completed by December, 1966. A defense date was set for Valentine's Day in 1967 and , therefore, several schools were interested in recruiting me. Ezra Stotland told me that a dissertation was simply a pedestrian effort to do research and I took him for his word. In the spring of 1966, Saxberg received a grant from Exxon to undertake an empirical study of ways students can improve their learning. We discussed various themes that focused on instructional pedagogy. From readings in social psychology, the concept of group cohesiveness and how a group affects an individual's performance was interesting. Within a few weeks, I drafted a proposal that was ultimately accepted by Exxon. During the summer of 1966, I formed a dissertation committee and submitted my formal dissertation proposal. The proposal detailed an experiment that would be carried out during the fall semester and would involve students taking introductory courses in the business school. Because Exxon and Kermit Hanson, the Dean, were sponsoring this research, I had access to student academic records and faculty teaching evaluations. In an experimental group, all students took their core classes together; in a control group, students classes were randomly scheduled. Controlling for faculty grading histories and student's past academic achievements, it was found that when students take all classes together they achieved higher grades, reported higher psychological adjustment to the school, and rated faculty members better instructors than those students in the control group. While the experiment was underway in the fall of 1966, the first three chapters were drafted and approved. Once the term was completed, the data analysis was undertaken and a discussion section written. Saxberg admonished me for not writing enough (I handed in only 75 pages), and wanted more. Within a week or so, I wrote ten more pages. He accepted the additional rhetoric and we scheduled a defense.

In early January, 1967, I interviewed at several midwestern schools, but the chemistry wasn't right for them or me. In late January, I arranged to interview at both Syracuse and Penn State. After a long interview at Syracuse, I flew into Phillipsburg, PA, an airport located in a state forest 25 miles from Penn State. There was a message from Max saying get a cab and call him after I had checked into the Nittany Lion Inn. Max had arranged a small party at his house with other faculty members. People were drinking, but since I hadn't eaten all day, I needed some food. His wife made me a peanut butter and jelly sandwich and the interview started. After spending some time at Penn State, I flew home. Max told me to call him before making any decision. Gail and I decided that Penn State would be the best place to start.

There were several reasons for this decision. First, Wally Hill, a management faculty member at Syracuse, advised me against joining him at Syracuse. He was not pleased with the school and was thinking about joining the faculty at Florida. Second, the weather in Syracuse was terrible—snow and cold—and reminded us of Seattle—drizzle and no sun. Third, I wanted to be challenged by the best management scholars in the country. At that time, Penn State's management faculty was among the very best. The opportunity to go toe-to-toe with the "big guns" was exciting. Lastly, Gail thought that she could get a teaching position in State College and was interested in pursuing her master's degree in reading at Penn State, which she completed in 1971.

I called Max and told him that I wanted an offer. He outlined an offer on the phone ($12,500 for nine months, teaching 9 hours a term for three out of four terms, and two preparations), at which time I told him that I had an offer from Syracuse. Days passed with no "formal" offer from Penn State. Late in the afternoon, I got a phone call from Max asking me why I didn't respond to his offer. I said that I never received it and was going to call Syracuse and accept their offer. Max said that Penn State had sent the telegram to our apartment, but since Gail and I weren't home, it was just tacked onto our apartment door. I walked to the apartment and got the telegram and called Max to say that I was going to accept Penn State's offer.

Throughout my career, I have made it a practice not to mix home and work. I have never had a home office, and still seldom grade papers or do any professional work at home. Therefore, this essay is partitioned into personal and research portions.

EARLY CAREER AT PENN STATE

The Penn State management faculty had been rated among the very best in the country in 1967. Rocco Carzo, John Yanouzas (J.Y.), Paul Greenlaw, and Max Richards had established national reputations in the management field. Rocco and J.Y. had just published their book *Formal Organizations*, and had a major article published in *Administrative Science Quarterly*. Richards and Greenlaw's book, *Management*, was in the second edition and Max was former President of the Academy of Management. I felt that it was a privilege just to work with these people. The pressure to earn their respect was tremendous. The *Management Science* acceptance occurred during that first winter and there was quite a celebration.

The Penn State faculty worked and played hard, especially at racquetball and golf. The competitive nature of these individual, as opposed to team sports, was consistent with a school culture that attracted an excellent cohort of young, aggressive management faculty members with strong research agendas, including Dick Chase, Mike Misshauk, and Gerry Susman. To compete and survive in such an arena demanded personal sacrifices. Faculty members worked seventy or more hours a week on their research projects. Every weekday night at 7:00 the junior people were back working in their office on research projects. Returning home before 9:30 or 10:00 violated the "informal" norm that had been established.

The teaching norm was survival. Teaching two business policy and one principles of management courses six days a week was a hardship. Classes were large and student/faculty interaction was minimal. Faculty members administered multiple-choice examinations that could be graded by the computer or a teaching assistant. Teaching assistants taught smaller discussion sections of the course and handled all student complaints.

Competition for scholarly recognition was also keen between junior faculty members across departments. Teaching ratings and publication records were constantly compared, along with salary increases. Ron Copeland, a member of the accounting department, compiled a list that ranked all "young Turks" from highest to lowest on their scholarly achievements. Research seminars held in the school were opportunities for junior faculty members to demonstrate their knowledge and penchant for methodological rigor. Ideas generated at these seminars were vigorously debated. After such heated debates, it was time for poker and beer. Carzo, Richards, Yanouzas played cards along with the junior faculty members from various departments. Poker games often lasted until the wee hours of the morning. On more than one occasion, I remember sitting at the New College Diner at 5:30 a.m. ordering breakfast and thinking about a forthcoming golf game at 7:00 a.m.

Max became my mentor. We shared a passion for golf and frequently played golf at Penn State, a sport I learned and honed from my father who was a good golfer. Max and I kept track of wins and losses and constantly needled each other about missed putts or other poorly played shots. Golf is played to win, and as Leo Durocher, former manager of the New York Giants, once said, "Nice guys finish last." Neither Max nor I wanted to be a member of the nice guy club. The competition on the golf course spread to spontaneous debates on any academic topic. It was through these activities that Max's philosophy of management and scholarship shaped my thinking about the profession. Max's repute as a teacher/scholar was renowned. He had an ability to quickly read a paper and reduce its contributions to a few sentences. Red marks and corrections dotted your returned manuscript. A few faculty members were intimidated by his demeanor and eschewed his help. Unfortunately, they never realized his brilliance. He also demanded classroom excellence from his students and the nickname "Max the Ax" stuck and was used by all. However, it was through these conversations and rich dialogues that I appreciated his contributions to my personal and professional development.

Learning from Max was challenging. Max stressed research on interesting problems that were theoretically relevant to the management field. His idea for relevant theory building was greatly influenced by Robert Dubin's book, *Theory Building*. Hours were spent on discussing whether a variable, such as climate or job satisfaction, was summative or not. According to Dubin and Max, summative variables should not be researched because they were theoretically vacuous. He advised young scholars to advance theory and not get embroiled in publishing "quick" methodological pieces that really did not contribute to the field, but often were published in the field's most prestigious journals. He decried faculty members who published a study that demonstrated the reliability of a scale, such as Porter and Lawler's scale to measure Maslow's needs hierarchy in the *Journal of Applied Psychology*. According to Max, these manuscripts contributed little to theory building. To date, I have only published one methodological article in my career, which ironically was published in the *Journal of Applied Psychology*. This article was a longitudinal analysis of employee's need deficiencies and how these deficiencies, as measured by the Porter and Lawler scale, affected both job satisfactions and performance.

Max was a superb teacher. He knew the concepts and was conversant with how a theory could be applied to solve administrative problems embedded in Harvard Business School Case Studies. He rarely relied on teaching notes, believing that faculty members should be completely prepared for class. Using the case method, he would drive, if possible, to the company and discuss the case with the company's managers. He averred that faculty members should understand first hand how an organization implemented solutions to its problems. He had a great capacity to memorize students' names and called on them in class. Students who were unprepared did not attend class that day.

In 1969, Penn State tried to recruit Don Hellriegel to join our faculty, but Don joined the faculty of the University of Colorado, Boulder. I was extremely disappointed with our inability to hire Don. From our decade long friendship, I had an innate sense that he would fit into the culture that Penn State had created. However, during Thanksgiving recess in 1970, Don and I got together in Cleveland. Don was visiting his family and I was visiting my wife Gail's family. He discussed his displeasure with some aspects of Boulder. Upon returning, I told Rocco, who quickly contacted Don. Don visited Penn State over Christmas vacation and we extended him an offer that he accepted.

During the winter months of 1971, Don convinced me that we should write a management textbook. Don thought that with the growing emphasis on the contingency approach, a textbook that featured this would be able to capture a significant share of the market. He also felt that there was not a significant amount of competition. The fragmented market was populated with traditional management books and younger faculty were tired of teaching "principles of management." With the growing enrollment in business schools, the financial returns looked promising. Outlines were exchanged and, to my surprise, we had multiple publishing houses wanting to publish this book. Offers to fly us to San Francisco to "think" about the text, etc. were made. Gail was pregnant with our second son, Bradley, and unfortunately couldn't fly anywhere. Max said forget the advances, trips, etc. and sign with the person who you trust and the company that could market the book. Applying these criteria eliminated several publishing houses. In the spring of 1971, we signed our first book with Addison-Wesley. Our editor was Dick Fenton.

Don's arrival at Penn State was a major event for me and the department. Between January 1 and Don's arrival in June, Rocco decided to leave Penn State for Temple University, and John Yanouzas joined the faculty at the University of Connecticut. Penn State was on the verge of losing its reputation as an academic powerhouse. With Don, we were able to recruit Chuck Snow and Larry Hrebiniak in 1972. These two scholars immediately fit into the culture of Penn State and were very productive. Although Chuck and Larry didn't have the status in the field that Rocco and J.Y. had earned, they were great colleagues. It was also at a time when Penn State attracted some excellent doctoral students, such as John Sheridan, Kirk Downey, Carl Anderson, Herb Hand, Bob Keller, Bill Joyce, Don Hambrick, several of whom are established scholars today.

Other schools actively tried to pull Don and I from Penn State. Because of budgetary problems and a lack of any opportunity to make money consulting or through executive management development programs, Steve Kerr tempted me to join the faculty at Ohio State in 1974. This was a personal challenge for Gail and me because we had just built our dream house for $45,000, and she was expecting our third child, Jonathan, in the summer of 1975. Penn State promoted me to full professor in 1974 but could not award me tenure because I lacked sufficient seniority. A sizable pay increase was used to placate my disappointment.

MID CAREER CRISIS

As fate would have it, no sooner had we agreed to go to OSU for 1975-1976, than Bob Strawser, a former Penn State accounting faculty member who was now at Texas A&M, called Don and me. Don was also growing restless about the financial situation at Penn State and took an interview trip to A&M in the winter of 1975. Upon returning to Penn State, he called and said he was leaving to go to A&M.

Our stay at Ohio State was wonderful. Ned Bowman was the dean and an excellent scholar who appreciated and rewarded scholarship. I remember giving Ned a draft of a paper that Hank Sims and I were writing. Within 24 hours, Ned had read the manuscript and given us detailed comments on ways to improve the manuscript. Steve Kerr, Randy Bobbitt, Chuck Behling and Ralph Stogdill were superb colleagues to exchange ideas with. My academic vitality was recharged.

Research seminars resembled Penn State's except that OSU was able to attract academic superstars such as Karl Weick, Henry Tosi, Larry Cummings,

Andy Van de Ven, Bob House, Bill Starbuck, to present their work to faculty and doctoral students. Jeff Ford, Janet Fulk, Chet Schriesheim, Mary Ann VonGlinow, Jody Fry, John Jermier were all doctoral students eager to show their academic prowess to the faculty and the visitors. The research collegia became forums where debates were commonplace among students, faculty members, and guests. The academic productivity of the department was great. In early 1976, Steve told me that he would be leaving OSU because he was divorcing his wife. The attraction to OSU waned considerably after that because Steve was a tremendous intellect, eclectic scholar, and was networked with the best thinkers in the field. His repute as a scholar was instrumental in attracting talented doctoral students to OSU's program. We had agreed to write a chapter in Starbuck's *Handbook of Organization Design* on groups with Ralph Stogdill. The chapter would coalesce Steve's knowledge of leadership, my knowledge of organizational design, and Ralph's knowledge of group processes. Steve and I had many meetings with Ralph who served as our teacher and directed our attention to group process literature that was foreign to us. Losing intimate contact with Steve because of his relocation to University of Southern California would strain our academic growth. Unfortunately, immediately after my return to Penn State and Steve's departure to USC, Ralph passed away. Steve and I somehow managed to write this chapter, but had to refocus it to capture what we knew.

From 1976 until 1978, the reputation of Penn State's management faculty grew as faculty members published in the field's most prestigious management and psychology journals. The core management research faculty now included Gerry Susman, Chuck Snow, Bob Pitts, Larry Hrebiniak, Max Richards, Hank Sims and Paul Greenlaw. Several of our doctoral students had been placed at Wharton, Maryland, Kentucky, Pitt, Indiana, and Houston, among others.

Within a short time, financial problems once again plagued Penn State. Don called me and told me about a chair position at Southern Methodist University (SMU) in the Cox School of Business. I had never thought about SMU and didn't know any management faculty members there. But after thinking about living in a warm climate and a major metropolitan area, Gail and I decided to take a recruiting trip. The trip to SMU was wonderful. Departing from the snow and arriving in the warm sunshine was quite nice. Bill Brueggeman, a former Ohio State faculty member, held a chair in real estate at SMU and sold me on how several people could really make an impact on SMU's scholarly reputation. The faculty were relaxed and involved in non-academic matters, such as golf and consulting. Alan Coleman, SMU's dean, embodied the characteristics of a successful businessman, as well as a person who understood and rewarded intellectual achievement. Gail and I thought that this might be a good move and waited for SMU's decision. Finally, in June, Alan called and offered me the Corley Chair and said that Mick McGill would present the formal offer. We decided to stay at Penn State for the 1978-1979 academic year and move to SMU in June, 1979.

Throughout the fall of 1978, I flew to Dallas several times. Meeting with faculty and executives, playing golf, and enjoying the weather reaffirmed that we made the correct decision. I also knew that SMU was not Penn State. There would be no more doctoral students or faculty members whose academic work consumed 70 hours a week. Cognitively knowing this, but then actually facing the fact would prove to be a major psychological hurdle.

Alan understood the "empowerment" process before it became a management fad. Faculty made most major hiring decisions and he always reserved time to listen to his faculty. Very quickly, however, SMU's faculty experienced "turnover," as key research faculty members gained appointments to the tenure and research & development committees. Junior faculty members who didn't have an active research agenda were counseled to leave and replaced with teachers/researchers. Alan was able to raise millions of dollars for endowments from prestigious local Dallas executives to create additional endowed chairs in all departments. Chairholders from prestigious schools, such as Wharton, Tuck, Chicago, Carnegie Mellon were hired. Norms quickly developed about publishing and many faculty developed an aggressive research agenda.

Alan resigned in 1981 and we began the search process for a new dean. The search committee reflected the emerging norms of the school concerning research and executive education. The Executive MBA program was SMU's flagship graduate program and the committee searched for a person who had leadership experience in that program, along with the ability to raise funds for new physical facilities. Roy Herberger from University of Southern California was hired and immediately developed a warm relationship with the business community and many faculty members. Roy was an avid golfer and we spent many Friday afternoons on the course nurturing relationships with local business leaders. Roy's charisma was instrumental in hiring high caliber junior colleagues, raising funds for two new buildings, renovating to the existing building, and attracting additional renowned faculty members in the MIS area.

Chairholders from the various colleges within SMU (about 55 in all) formed an eating club and met several times a semester to exchange scholarly perspectives. My appointment to several prominent all-university wide committees on athletics, fund raising, and scholarship enabled me to gain a broader perspective of SMU than I had been able to gain from the business school. Access to the president and provost also satisfied my need to "make things" happen. Roy also appointed me to key administrative committees within the business school. Coterminous with this increasing "influence" at SMU was my election as 39th President of the Academy of Management.

I started a sabbatical leave in the spring semester of 1989. In February, Roy abruptly quit SMU and took the job as President of the American Graduate

School of Management. The Provost asked me to become dean ad-interim while a search committee was appointed. After thinking about it and talking it over with Gail, I accepted the challenge. Gail and I had planned a two-week trip to France, but I knew that the demands of the office wouldn't permit this trip to take place. Canceling the trip was another personal/professional tradeoff.

The dean's job provided me with a valuable learning vehicle to closely examine my administrative strengths and weaknesses. Learning how to influence the provost to give extraordinary salary increases to deserving faculty, working with American Association of Collegiate Schools of Business to gain reaccreditation for the school, establishing and capitalizing on donor relations that provided scholarships for undergraduates and negotiating with Ken Pye, SMU's president, were welcomed learning spots. Trying to "herd" faculty members to work on priorities other than theirs, integrating curriculum issues across colleges, and providing academic leadership were scintillating experiences. Also, I quickly learned how "lonely" the job was. You could share feelings and thoughts with only a few people. It was difficult to maintain an active dialogue with departmental and other colleagues because they treated you as an administrator. One associate dean left to go into private practice, the controller retired and my chief fund raiser joined Roy at Thunderbird. Although they left for career reasons, suddenly the astute administrative staff assembled by Roy and with whom I had worked for years was gone. New relationships had to be quickly established.

My most cherished dialogues occurred with Ken Pye. Ken, a lawyer by training, had many strong beliefs and conveyed these to deans, donors, faculty members and students. I remember one of our Friday staff meetings during which Ken announced that by next Tuesday, he would like to see a list of what undergraduate classes the chairholders would be teaching next fall. It was members of SMU's Board of Governors and Ken's belief that all endowed chairholders should be required to teach undergraduates. This wasn't a debatable subject. A few of us scurried to the bar for lunch and tried hard to figure out how to break this news to chairholders, many of whom had not taught an undergraduate class for years and were attracted to SMU because they would have no undergraduate responsibilities. After lunch, I called a meeting with the departmental chairpersons and chairholders and told them of Ken's decision. Open resistance and a questioning of my administrative abilities were discussed. However, by Tuesday, all business school chairholders had been scheduled. The lesson I learned from Ken was that he had to make a difficult decision that would probably be resisted by all parties. Discussion and debate wouldn't change his position. Just make the decision and move on.

Ken and I had a similar discussion on the vacant dean's position. Ken wanted (demanded) the Cox School to drop our one-year MBA program and adopt a traditional two-year program. His argument was that if better schools, Duke,

Harvard, Tuck, Wharton, required MBAs to enroll for two years, why shouldn't SMU. Given the financial exigencies facing the business school and the university because of the football scandal, I couldn't justify and commit the school to this program. Undergraduate enrollment at SMU had fallen considerably as a result of the National Collegiate Athletic Association banning football at SMU for one year and then SMU voluntarily not playing another because of illegal recruiting practices at SMU. Because we couldn't agree on the strategic direction of the school, I declined to become a candidate for the dean's position. While attending the annual Academy of Management meetings in Washington, DC, Ken called to tell me that David Blake was going to be the school's new dean. The learning lesson was that unless you can commit yourself to the goals and objectives of the president, you shouldn't occupy an important administrative post.

Blake arrived in January, 1990. Within a few hours after returning to my faculty office, the phone stopped ringing. Pressing administrative needs, however trivial, that consumed hours of time, were now being addressed by someone else. I immediately began my sabbatical and Gail and I spent time away from SMU and Dallas. We took another sabbatical two years after that to go to Japan so I could teach at the International University of Japan.

LATE CAREER

I have always thought of myself as an outstanding teacher. I set goals to win SMU's two most prestigious undergraduate teaching awards and be recognized as a sought after person in executive development programs. A total immersion into the undergraduate curriculum was required, along with a total dedication to this program. Within two years, I had won both the school and university wide teaching awards and have since won the school's best teaching award again. Only Mick McGill, a colleague in my department, has won this award twice. I found that working in many university executive development programs, most notably Penn State's, and a host of organizationally sponsored programs, provided intrinsic rewards that were not forthcoming from the school's current administration. In contrast to Roy's warm and outgoing interpersonal style that fostered a supportive and "fun" place to work, the administrative posture now in the business school has sent me to look for rewards outside of the Cox school.

Through active recruiting of research/teaching faculty, the external reputation of the Cox School, and in particular the management department's, has dramatically risen. The research output of the department is now more continuous and visible. Colleagues are addressing salient theoretical issues in the field and have aspirations to publish in the field's most prestigious academic and practitioner journals.

Reflecting on my administrative experiences, I have looked at several dean's positions. However, these positions required that we move from Dallas, a city that we love. The demands of a deanship also conflict with my desire for independence and opportunity to participate in executive development programs throughout the world. These considerations led to a decision to remain at SMU in a faculty, as opposed to administrative, position and work on SMU administrative activities external to the Cox school. I have developed a passion for teaching that continuously renews and challenges me.

EARLY PROFESSIONAL CAREER

The culture at Penn State was supportive for doing high quality creative research in a myriad of disciplines. Some of my early research focused on credit card usage patterns and was published in the *Journal of Marketing*. These were seminal studies and widely acclaimed in the marketing literature. As my departmental chairperson, Rocco reminded me that if I chose to publish outside of the major management journals, the work should reflect the highest standards of excellence that could be achieved in other fields. Max, on the other hand, told me that I needed to dedicate myself to the management field. I've largely followed Max's advice and focused my attention on research projects that would enable me to publish in management.

My penchant for doing eclectic research attracted my first doctoral student, Herb Hand. Because I was an untenured faculty member, I couldn't direct Herb's dissertation. Max signed on as Herb's official advisor and we co-directed Herb's dissertation. From that study, a stream of research that focused on climate, job performance, and work-related attitudes started. There was no underlying theoretical focus to this research, except that it was being continuously published in the most prestigious management journals and satisfied my intellectual curiosity. It also afforded me the opportunity to get promoted to an associate professor within two years after arriving at Penn State.

In 1971, several events occurred that were instrumental in shaping my career. First, a goal to become president of the Academy of Management was set. Max indicated that with my academic visibility and Penn State's reputation, I could become president of the Academy of Management if a clear set of goals were formulated. To achieve this goal, he needed that I become actively involved with the political networks of the Academy. Becoming the president of the Eastern Academy of Management (EAM) was politically expedient because of Penn State's reputation in the eastern part of the country. I was able to get elected to the EAM Board of Governors, serve as program chairperson, and ultimately serve as president, a feat accomplished in 1974. To expand my arena of academic influence, membership on the *Academy of Management Journal's* (AMJ) editorial review board was required, as well as becoming actively involved in the divisional structure of the Academy of Management. Second, Penn State had attracted some outstanding doctoral students, namely Kirk Downey, John Sheridan, Bob Keller, and Don Hambrick. Having the ability to work with these students greatly enhanced my ability to publish research in top-tier journals because each was pursuing different lines of research. Attracting talented Ph.D. cohorts who could cooperate in joint research projects was critical.

ACADEMY YEARS (1972-1985)

Jack Miner was in the audience when I presented a paper at an Eastern Academy of Management meeting in 1972 and was impressed with my knowledge and penchant for methodological rigor. He asked to me join his editorial review board at AMJ in 1973. Working with Jack exposed me to a wide range of theoretical and methodological issues, because Jack believed that all board members should process a wide diversity of manuscripts. It also afforded me the opportunity to learn about the Academy and some of its challenges.

In 1972, I was elected program chairperson for the Organizational Behavior (OB) Division of the Academy. This position enabled me to gain keen insight into the strategic decision making processes of the Academy and establish a network with a host of colleagues who governed the Academy. While chairperson of the Organizational Behavior division, the division decided to sponsor a doctoral consortium prior to the Academy's annual meetings being held in Kansas City. Modeled after the American Marketing Association's doctoral consortium program, the OB division hosted the Academy's first consortium without the financial support of the Academy. It was an immediate success and the next year was co-sponsored by the Organization and Management Theory (OMT) division. A year later, I was elected program chairperson for the Organization Theory Division and was in line to become president of both divisions in the same year. Although this accomplishment didn't violate either the bylaws or constitution of the Academy, Stan Vance, president of the Academy, asked that I step down as president-elect for the OMT division. Naturally, I was disappointed with his request, but after a long discussion with Max, Larry Cummings, Andre Delbecq, I stepped down as president of that division. That request and my decision still haunt me today.

In 1976, I was elected an Academy Fellow and to the Board of Governors. Larry Cummings was elected editor of *Academy of Management Journal* and asked me to serve on his editorial review board. The Board had just approved the creation of the *Academy of Management Review* (AMR) and now turned its attention to governance issues facing the Academy. The creation of new divisions and interest groups, active recruiting of new members, and the dominance of the "O" divisions (Organization and Management Theory and

Organization Behavior) in Academy affairs dominated the Board's discussion. Past presidents and journal editors had all held powerful positions in the "O" divisions and dominated the composition of the Board of Governors. The then American Institute for Decision Sciences (AIDS) and now Decision Science Institute, a name change made for obvious reasons, made a faint hearted effort to attract the OB members by forming a new division in their organization and accepting organizational behavior manuscripts. Because AIDS appealed primarily to operations management and decision scientists, many of the Academy's members did not look to this organization as their academic home. Fortunately for the Academy, those of us advocating growth persuaded others that membership growth was necessary for the professional and financial viability of the Academy. New divisions were created and interest groups formed as the diversity of the Academy's membership grew by primarily attracting faculty members in business policy and industrial psychology.

In 1978, Larry asked me to become his Associate Editor with the understanding that I become editor of the *Academy of Management Journal* in 1979-1981. The editor was not only in charge of selecting editorial review members to determine the suitability of manuscripts to publish, but also negotiated with a publisher to print the journal, and solicited advertising and new library subscriptions. AMJ and AMR were run as profit-centers and were expected at least to break-even. The host institution of the editor was expected to provide office space, administrative assistants, postage, and the like. I hired Barbara Williams, Jack Miner's administrative assistant at Georgia State and who later became his wife, to work with me via long distance in Atlanta. Barbara negotiated rates with a local Atlanta publisher and the post office, and secured the editorial assistance of Mary Bowdoin to copy edit the journal.

Unless you have actually been there and done it, it's impossible to know the demands and frustrations of being an editor of a major journal. You cannot wait for instructions from somebody else. Personal initiative is essential for survival. Thirty or more hours a week more were common and when we moved the journal's editorial operations from Penn State to SMU during June of 1979, it was a lot more. I found that self-discipline was needed. Disappointed authors, publisher demands, and post office problems can burn an editor out. During my tenure as editor, I processed more than 1,174 manuscripts and published 174. I spent countless hours reading reviews and writing acceptance and rejection letters. A conscious attempt was made to attract more business policy manuscripts to the journal by adding reviewers in this field and publishing articles that posed interesting research questions. During my last year as Editor, I became a candidate for the Academy's vice-president program and was elected to that position in 1981. Because I had served on the Board of Governors for years, I understood the demands of this new position and relished the opportunity to pass on the editorial baton to Tom Mahoney. I ran the 1982 Academy meeting in New York City, when

Max Wortman was President. It was a joy to work with Max and I relished his easy going attitude.

Overseeing a national meeting was a huge managerial undertaking. Besides focusing on divisional concerns for time and space, working with the on-site hotel's professional staff and unionized employees, publishers, advertising agencies, and the post office, the meeting was being held in a city 1350 miles away from Dallas. It tested my ability to share decision making to others and manage results, not activities. The ability to empower others and delegate key decisions proved to be difficult for me because my natural tendency is to "take charge and get things done." The frustrations of working with people who promise everything and produce nothing strained my leadership competencies.

The composition of the Academy's Board was eclectic and tested my administrative acumen. As president-elect, I proposed that the Academy create a new journal that would appeal to professional practitioners and to the teachers in the Academy. I was increasingly concerned that both AMJ and AMR were publishing articles that didn't appeal to many of our members and couldn't be used in the classroom. Task forces were asked to write "white papers" on the merits and pitfalls of this new journal. After much heated debate, and some strong-arm tactics being applied to Jan Beyer and Jeff Pfeffer, the Board approved this new journal during my presidential year. Under the astute presidential leadership of Fred Luthans and Kay Bartol, the Board implemented this decision and selected Warner Burke to be the new journal's first editor. The joy and excitement of seeing the first edition of the *Academy of Management Executive* in March, 1987, was tremendous. Finally after five years, my vision for this journal had culminated. Warner asked me come on board as his associate editor, which I did with the full knowledge that I would not seek nor accept the journal's editorship. While later on in my career I will hold other editorial positions with *Organization Science, Decision Sciences*, and *Organizational Dynamics*, the sense of accomplishment has never been greater than with seeing the first issue of the *Executive*.

From holding various leadership positions in the Academy, I learned the importance of establishing a clear vision of where you and your associates want to take an organization. Getting ideas out of your head and into the heads of others takes excellent communication skills. This meant listening to Jan Beyer's objections to the *Academy of Management Executive* while sitting in a hot tub in San Diego after a Board of Governors meeting. It also means taking a lot of fuzzy information, drawing implications, and trying to figure out what all the stuff means. While not everybody will be as "upbeat" as you are in implementing new ideas, you need to reinforce people for taking steps into uncharted waters with you. Cajoling dissenters while being able to make things happen is critical. Taking the *Academy of Management Executive* from the idea stage to its actual production was exhilarating. Giving credit to others who disagree with your vision is difficult, but must be done if you desire teamwork. In essence, I learned that the goal of every person is to be recognized as a valued member of the team, regardless of whether those people agreed with the decisions of the majority. Being respectful of others and their different points of view was hard to learn. Decisions that needed to be made created untold hours of debate and tension. Much self-discipline is necessary to lead an organization. That means one must be able to set priorities and manage their own time. Editors and program chairpersons must also learn to feast off their own sense of self- accomplishment and shrug off defeats. Few colleagues will take the time to say "thanks," or even drop you a note of appreciation. Finally, I got to know myself and my strengths and weaknesses.

RESEARCH AGENDAS

The ability for a school to accept and utilize my widely differing research contributions has been an integral part of my research portfolio. With the arrival of Don Hellriegel at Penn State, the beginning of our quarter of a century textbook writing career started. Our first book, *Management: A Contingency Approach*, was published in 1974 and this was followed quickly by the publication of *Organizational Behavior* in 1976. Both books are now in their 7th editions and have found broad acceptance in the academic marketplace. Writing these books has greatly improved my instructional effectiveness because they have kept me current with a large body of practitioner literature, including *Fortune, Inc., Forbes*, and *Wall Street Journal*, that I would otherwise not read. The ability to translate theories into useful knowledge that students can understand has been a major challenge. Don and I are able to tell each other the "truth;" that is, give each other brutal comments. Heated debates often take place, but a resolution has always been reached. We have unsuccessfully tried to "outsource" selected chapters to various colleagues, but these attempts have largely failed. The exacting standards we have followed for 25 years, along with our ability to find creative solutions to problems have been difficult to translate to others. They usually lack the courage to stand up to the potshots delivered by one of us or other reviewers. We have found out that it frequently takes "hands-on" editing and commentary to make a real difference in a manuscript. We have also found that other faculty members have failed to keep up with academic literatures or they have become purveyors of management myths. Exceptional writing ability is vital.

Don and I have also been co-editors for West Publishing Company's Management Series. In a span of eleven years, we sponsored more than twenty books, most notably Dick Daft's *Organization Theory and Design* book that is now in its fifth edition and Randy Schuler's *Personnel and Human Resources Management*, which is also in its fifth edition. This experience afforded me tremendous insights into the publishing industry, especially its cost structure and marketing strategies.

Our success in the textbook field has never been translated into research publications in scholarly journals. The fact that we are both strong conceptually and lack strong statistical skills might be one explanation. We have written several professional practitioner manuscripts that creatively synthesize our ideas on organization design and problem-solving skills of executives, but most of my earlier research was written for students. Furthermore, we have not been on the same faculty together since 1975. Empirical research requires reciprocal task interdependence and we have largely used pooled interdependence to achieve our textbook writing goals.

Blessed with superior doctoral students at Penn State and Ohio State, I was able to pursue multiple research agendas. One research agenda focused on organization theory topics. Along with Kirk Downey, Jody Fry, Bob Keller, John Jermier, and Jeff Ford, a series of research questions that focused on organization design issues ranging from environmental uncertainty to technological determinism were studied. I relished the opportunity to read different literatures and coalesce these into integrative wholes. The outputs of these research projects were published in *Administrative Science Quarterly, Academy of Management Journal, Human Relations, Academy of Management Review, Organizational Behavior and Human Behavior, Journal of Applied Psychology*, and other leading journals. All of these people, but primarily Kirk, had a tremendous appetite for work. As a result, he and I were able to publish nine articles and one organizational readings book together in a span of ten years.

Serendipity also played a role in my ability to craft numerous research ideas and get them published. Jeff Ford and I crafted our article on "Size, Technology and Environmental Uncertainty" while driving from Columbus, Ohio to Cincinnati to deliver a paper on leadership at a Decision Science Conference in 1976. Upon returning to Ohio State, we quickly drafted a manuscript and forwarded it to the *Academy of Management Review*. The reviewers' comments were mostly editorial, as opposed to substantive, and this article was quickly accepted. This article has been widely cited for its ability to creatively synthesize divergent literatures and pose new directions for the field.

As Kirk was graduating from Penn State's program, John Sheridan enrolled. An engineer by training, John brought methodological sophistication to our research team. We crafted a series of longitudinal studies that focused on expectancy theory and leadership using panel research designs. We read David Kenny's book, *Correlation and Causality*, and used that methodology to investigate a myriad of topics. Recently, with John's knowledge of survival analysis and my knowledge of career paths, we completed a longitudinal study of managerial careers in a large public utility. Over the past twenty years, John and I have pursued numerous research questions and have published 11 articles together in the *Journal of Applied Psychology, Academy of Management Journal, Organizational Behavior and Human Performance,* among others. Just like Don and I, we have always been willing to share constructive criticisms and work to resolve differences. We also have been consultants for numerous organizations and remain close personal friends as well as good colleagues.

When John Sheridan was about to graduate, Bill Joyce walked into my office. As a masters student at Bucknell, he needed some literature on matrix management to complete his Master's thesis. Bill was an aerospace engineer working with Piper Aircraft in Lock Haven, PA. I convinced Bill that he should enroll in Penn State's doctoral program, which he did. Bill's thinking was greatly influenced by the research of Ben Schneider and Larry James. Bill was able to interest me in the climate area and we crafted several studies that linked climate perceptions to performance and the formation of climates to various design characteristics of organizations. Bill was able to use his analytical skills and his interest in the philosophy of science to develop sophisticated mathematical models of organizational climates. Our work in this area has been published in the *Academy of Management Journal, Human Relations, Academy of Management Review*, among other journals. We have also become lasting friends and spent one semester together in Japan with our wives. It's a semester that will not soon be forgotten. During that semester, we initiated a project that focuses on attempting to make sense of the corporate diversification literature using sophisticated mathematical modeling techniques that we have been learning from a colleague of Bill's at Dartmouth. As usual, this project will take years to complete.

RESEARCH AGENDA: MID-CAREER

Once the decision to join SMU's faculty was made, I realized for the first time that there would be no doctoral students to maintain by eclecticism. However, through a random meeting after an academic conference at the University of Texas—Arlington, Bill Cron and I discovered that we had a joint interest in career issues. We carefully crafted a series of studies that explored career paths of salespersons in organizations that were pursuing various business strategies in different industries. Just like others who I have worked with throughout my career, Bill had a tremendous appetite for work and had excellent statistical training. Our work on careers has been published in the *Academy of Management Journal, Journal of Marketing Research, Journal of Vocational Behavior*, among others. Bill and I are presently working on research that focuses on the performance of retail stores and another study that examines the impact of goal-setting on the performance of salespeople during a promotional campaign.

In 1990, David Lei was recruited from the University of Texas, Dallas to SMU. David is perhaps the most eclectic scholar I've ever known. His appetite for work is unparalleled. David asked me to read a draft of a completed manuscript when I was the dean of the Cox School. Within a day, I had made comments on the manuscript and returned it to him. In reading this work, I was struck by how carefully he was able to articulate his arguments and how creatively he used examples from the practitioner literature to emphasize salient points. We immediately decided to write papers together that would combine his intimate knowledge of business strategy and technology and my knowledge of organization design. Because both of us are broad-gauge thinkers and have far reaching intellectual interests, we have researched and published work on joint ventures, reward systems, and learning organizations. There is little coherence to our work. Most of our work has been targeted for the professional practitioner and has been published in *Organizational Dynamics* and *California Management Review*. As with all others with whom I have collaborated, interpersonal attraction, rather than discipline orientation, provides the vehicle for initiating and sustaining our joint work.

RESEARCH AGENDA: LATE CAREER

During a meeting of executives at INSEAD in 1991, my thinking on organization design was influenced by Clyde Dilloway, Vice President of Organization Development at British Petroleum. As a discussant for how learning organizations employ learning concepts, I found Clyde's observations penetrating. Upon returning to SMU, I told Mick McGill of my intentions to write a professional book on designing learning organizations. Having read Peter Senge's book, *The Fifth Discipline*, and heard him lecture several times, I thought that a managerial audience seemed ready for this book. Unknown to me, Mick, who had written three other popular professional books, was preparing to write another one. After quickly comparing outlines, we decided to write a book on designing learning organizations targeted for the practicing manager. During the spring of 1992 while in Japan I was able to put some additional thoughts onto paper. Upon my return, we wrote several manuscripts on designing learning organizations that were published in *Organizational Dynamics*. We were able to work rather rapidly and constantly shared war stories with each other to highlight important concepts. David Lei also heavily influenced our thinking and gave us numerous examples. As a result, *The Smarter Organization* was published in 1994. The publication of this book established a vehicle for discussing problems with managers attending executive management development programs around the globe. We are presently considering writing another book with a Dallas executive on how conversations with customers, suppliers, managers, and government leaders affect learning and change in organizations.

During the next decade, I will make a tradeoff between being a productive academic scholar and translating useful knowledge for the professional practitioner. I find the intellectual stimulation of crafting, implementing, and disseminating information to my academic colleagues challenging, but not fulfilling. Our journals are replete with articles that have little impact on managerial practice. Producing useful manuscripts that can advance the status of managerial thinking is now a dominant part of my research agenda. Working on projects with Bill Cron, David Lei, Bill Joyce and others will satisfy my intellectual curiosity. As Don and I start thinking about the 8th editions of our the management and organizational behavior books, it is getting more difficult to keep the same level of enthusiasm for these books that we generated during earlier editions. Battling with reviewers, editors, and each other to translate exciting new managerial practices into concepts and language understood by undergraduates will be challenging.

SOME CONCLUDING THOUGHTS

Throughout my career, I have been blessed to be surrounded by a family and colleagues who not only have tolerated my idiosyncrasies, but who also share a passion for asking interesting questions. Don Hellriegel, Kirk Downey, John Sheridan, Bill Joyce, Bill Cron, Mick McGill, and David Lei have provided me with opportunities to make a myriad of contributions to the literature, as well my personal development.

The lack of a coherent research theme has permitted me to publish in a variety of journals and exposed me to a vast array of literatures. Teaching interesting courses that focus on managerial problems affords me the opportunity to continuously improve my instructional effectiveness and learn. My passion for teaching excellence is continuously heightened by preparing new materials for each course, even though the course title and number remains the same. This penchant for intellectual diversity has also been transferred to working in executive development programs. Receiving feedback from senior managers after designing and presenting new material has been personally rewarding.

Setting and achieving goals has been an instrumental part of my life. What are some of the more challenging goals that are now the focus of my attention? First, to become recognized as a truly gifted instructor. Student and executive evaluations indicate that I can clearly disseminate information in a scintillating manner. Colleagues frequently sit in on my classes to learn how to improve their instructional effectiveness. Observing others who have achieved instructional distinctions so I can learn from them is important. Learning from others occurs most easily when the person is not presenting topics that are familiar to me. This affords me the opportunity to watch the "process" of teaching and student learning, as opposed to thinking about the content. To reach a higher level of instructional effectiveness will require developing a passion for new materials that can be translated into the classroom. The development of new materials can be fostered by maintaining intimate relationships with clients and students. Gail and I frequently have undergraduate students over to our home for dinner. Listening to their concerns provides a vehicle to change classroom procedures. For example, permitting students to take alternative forms of exams, such as all multiple-choice, all essay, and oral, has greatly reduced the number of complaints about examinations. The ability to play golf affords me the opportunity to network with SMU's Executive MBA students and clients to better understand their needs.

Second, the mentoring of colleagues. I frequently read colleagues' papers and offer them critical commentary. The success of former students who have established their own reputation as scholars and administrators has brought tremendous personal satisfactions. The adage "by your subordinates' works ye shall be known" is important. The ability to feast off their successes has been personally rewarding. Gaining pleasure from the academic accolades received by colleagues, such as Bill Cron, David Lei, Ellen Jackofsky, Bill Joyce, Robin Pinkley, John Sheridan, and Gordon Walker, has been very rewarding. Working with Mick McGill has forced me to think outside of the academic box and , as such, I have learned to appreciate his contributions to my professional development.

Third, recognizing that I have reached all of my professional goals, such as being president of the Academy of Management, editor of the *Academy of Management Journal*, and being elected to Fellow status in Decision Sciences Institute and the Pan-Pacific Management Association early in my career has resulted in some frustrations. All of my professional goals were reached prior to being 45-years old. The ad-interim deanship at SMU and the ability to work with former president Ken Pye has teased my administrative aspirations. I need a vehicle by which I can productively channel these administrative energies and talents. Gail and I have just completed our second dream home in far north Dallas. Because we desire to reside in Dallas, my administrative energies will be channeled toward working on various SMU related activities, such as being the chairperson of the University Benefits Council, or a member of the Business Services Advisory Council.

Fourth, Gail and I have raised three great boys. Chris has received his Master's Degree in Social Work, is married, and working in Dallas with battered children and women. His has aspirations to return to the University of Texas and start working on his Ph.D. in social work. He and his wife, Haven, live very close to us and we see them quite often. Bradley is training to become a jet pilot for the U.S. Navy. He was commissioned and married to Christy shortly after his graduation from the University of Texas. Like his older brother, they are both die-hard "burnt-orange" fans which has led to some interesting debates around U.T.'s athletic achievements versus SMU's. Our youngest, Jonathan, is currently enrolled as a second-year student at SMU. He's majoring in English and hopes someday to write the great American novel. All three of our sons have played on city championship soccer teams and won numerous athletic awards during high school. As an athlete during high school, I regretted that my father never saw me play football or wrestle. Therefore, Gail and I spent countless hours on either freezing or hot soccer fields, transporting them to away games, postponing Thanksgiving dinner so the kids could go to San Antonio, Houston, or wherever to play in games. The only game we missed was when Chris was playing and incurred a major knee injury. I was at Penn State and Gail was doing volunteer work. To this day, I remember Gail calling me and telling me what happened. It was a sickening feeling not to be there when Chris needed us.

Fifth, to maintain and grow my association with good friends apart from SMU and Penn State. Fred Luthans and Sang Lee are avid golfers and we have played all over the world. Naturally, we keep track of wins and losses. Continuing to be a part of the Nebraska network and Pan-Pacific Management Association will afford me the opportunity to travel in Asia and learn more about management practices in that part of the world.

Finally, my golf game needs work and I plan to continue playing competitively several times a week. Wins and losses are duly recorded and money frequently exchanges hands. We belong to a world class country club and I presently sport a single digit handicap. Playing on week-ends with non-SMU people is important for my mental health. The country club is a refuge from the frustrations of lecturing, writing (and sometimes being rejected), and the politics of the Cox School.

ACKNOWLEDGMENT

The author thanks Ellen Jackofsky and Mick McGill for reading this epistle and making numerous suggestions. My life at SMU has been greatly enriched through my association with both of them.

PUBLICATIONS

Articles

1968

With B. Saxberg. The management of scientific manpower. *Management Science, 14*(8), B-472-489.

Sensitivity and self-awareness changes. *Journal of Training and Development, 22*(9).
Group cohesiveness: A salient factor affecting students' academic achievement in a collegiate environment. *International Journal of Educational Sciences, 2*(3).
With H. Mathews. *Marketing research strategies in the commercial bank credit card field* (82 pp.). Chicago: Bank Public Relations and Marketing Association.

1969

With H. Mathews. Social class and commercial bank credit card usage. *Journal of Marketing, 33*(1). [Reprinted in Kollat, Blackwell, & Engel (Eds.), *Research in consumer behavior*. New York: Holt, Rinehart and Winston, 1970; Kassargian & Robertson (Eds.), *Perspectives in consumer behavior*. Glenview, IL: Scott, Foresmen, 1973; Bliss (Ed.), *Readings in consumer behavior*. Boston: Allyn Bacon, Inc., 1973].
With M. Misshauk. Operative employees' satisfaction with selected work factors. *Personnel Journal, 48*(3).

1970

With M. Misshauk. Job satisfaction and productivity. *Personnel Administration, 33*(3).
With H. Mathews. Analysis of the bank credit card field by life cycle. *Journal of Consumer Credit, 1*(4).
The relationship of bases of supervisory power to employee satisfaction. *Personnel Journal, 49*(3).
Performance and satisfaction: An analysis. *Industrial Relations, 9*(4).
With H. Mathews. Social class and income as indicators of consumer credit behavior. *Journal of Marketing, 3*(2).
With H. Hand. Human relations training program: A field experiment in middle management. *Academy of Management Journal, 13*(4).
With R. Strawser. The impact of job level, geographical location and organizational size on managerial satisfaction. *Journal of Bank Research, 1*(3).

1971

With P. Adams. Group cohesiveness and employee satisfaction. *Personnel Administration, 34*(2).
A comparative study of United States and Mexican operatives. *Academy of Management Journal, 14*(1).

With D. Kuhn & R. Chase. The relationship between job performance and job satisfaction for operative employees. *Personnel Journal, 50*(6).

With P. Topichak & D. Kuhn. A cross-cultural study of need satisfaction and need importance for operative employees. *Personnel Psychology, 24*(3).

Motivation in managerial levels: The impact of job performance on satisfaction. *Journal of Applied Psychology, 55*(3).

1972

With R. Marsteller. The prediction of psychological need satisfaction from the Bernreuter Personality Inventory and biographical data. *Journal of Training and Development, 26*(2).

With R. Strawser. Racial differences in job satisfaction. *Journal of Applied Psychology, 56*(2).

With & P. Topichak. Does culture affect job satisfaction? *Journal of Applied Psychology, 56*(2).

With H. Hand. A longitudinal study of the impact of a human relations training program on organizational effectiveness. *Journal of Applied Psychology, 56*(5).

With C. Altimus & M. Richards. A cross-cultural study of need deficiencies of blue-collar employees. *Quarterly Journal of Management Development, 4*(2).

With H. Hand & M. Richards. Organizational climate and the effectiveness of a human relations training program. *Academy of Management Journal, 16*(2).

With H. Mathews. Social class and income: A note. *Journal of Marketing, 36*(4).

With G. Susman & J. Sheridan. An analysis of need satisfaction and job performance among professional and para-professional hospital personnel. *Journal of Nursing Research, 21*(4).

With D. Hellriegel. Systems strategies applied to organizations. *Business Horizons, 15*(2).

With H. Mathews. Correlates of commercial bank credit card use. *Journal of Bank Research, 2*(4).

[Editor] *Research in organizations* (185 pp.). University Park, PA: Proceedings of the Ninth Annual Meeting, Eastern Academy of Management, Boston, MA.

1973

With J. Sheridan, M. Richards, & C. Altimus. The impact of leadership styles on the job satisfactions of blue-collar workers in different cultures. *Quarterly Journal of Management Development, 4.*

With D. Hellriegel. A contingency approach to organizational design. *Business Horizons, 16*(2).

1974

With D. Hellriegel. Organizational climate: Measures, research and contingencies. *Academy of Management Journal, 17*, 255-280.

With R. Keller & G. Susman. Management system, uncertainty and continuous process technology. *Academy of Management Journal, 17*, 255-280.

With J. Sheridan & M. Richards. Expectancy theory as a lead indicator of job performance and job satisfaction. *Decision Sciences, 5*, 507-522.

With J. Sheridan & M. Richards. A comparative analysis of expectancy and heuristic models of decision behavior. *Journal of Applied Psychology, 60*, 361-368.

With K. Downey, D. Hellriegel, & M. Phelps. Organizational climate and job satisfaction: A comparative analysis. *Journal of Business Research, 2*, 233-248.

With D. Hellriegel. *Management: A contingency approach*. Reading, MA: Addison-Wesley Publishing Company.

1975

With K. Downey & J. Sheridan. Analysis of relationships among leader behavior, subordinate job performance and satisfaction: A path-goal approach. *Academy of Management Journal, 18*, 253-262.

With K. Downey & D. Hellriegel. Congruence between individual needs, organizational climate, job satisfaction and performance. *Academy of Management Journal, 18*, 149-155.

With K. Brown. Systems analyses of organizations. In D. Hellriegel & J. Slocum (Eds.), *Management in the world today* (pp. 54-70). Reading, MA: Addison-Wesley.

With K. Downey. Uncertainty: Measurers, research and sources of variation. *Academy of Management Journal, 18*, 562-572.

With J. Sheridan. The direction of the causal relationship between job satisfaction and work performance. *Organizational Behavior and Human Performance, 14*, 159-172.

With J. Sheridan & B. Min. Motivational determinants of job performance. *Journal of Applied Psychology, 60*, 119-121.

With D. Hellriegel & J. Slocum. Managerial problem solving styles. *Business Horizons, 6*, 29-37.

With K. Downey & D. Hellriegel. Environmental uncertainty: The construct and its operationalization. *Administrative Science Quarterly, 20*, 613-629.

Dimensions of participation in managerial decision making. *Atlantic Economic Review*, May-June, 32-35.

With J. Sheridan & K. Downey. Testing causal relationships of the path-goal theory of leadership effectiveness. *Organization and Administrative Sciences, 6*, 61-80.

With D. Hellriegel. *Management in the world today*. Reading, MA: Addison-Wesley Publishing Company.

1976

With W. Liddell. The effects of individual role compatibility upon group performance. *Academy of Management Journal, 19*, 413-426.

With K. Downey & J. Sheridan. The path-goal theory of leadership: A longitudinal analysis. *Organizational Behavior and Human Performance, 16*, 156-176.

With D. Hellriegel. *Organizational behavior: A contingency approach*. St. Paul, MN: West Publishing Company.

1977

With K. Downey & D. Hellriegel. Individual characteristics as sources of perceived uncertainty. *Human Relations, 30*, 161-174.

With C. Anderson & D. Hellriegel. Managerial response to environmentally induced stress. *Academy of Management Journal, 20*, 260-272.

With J. Ford. Size, technology, environment and the structure of organizations. *Academy of Management Review, 2*, 561-575.

With J. Sheridan. Causal inferences in motivation research: A reinterpretation of panel study results. *Journal of Applied Psychology, 62*, 510-514.

With W. Joyce & M. Abelson. A causal analysis of psychological climate and leader behavior. *Journal of Business Research, 5*, 261-273.

With K. Downey & D. Hellriegel. *Readings in organizational behavior: Contingency views*. St. Paul, MN: West Publishing Company.

1978

Does cognitive style affect diagnosis and intervention strategies of change agents? *Group and Organization Studies, 2*, 199-210.

With D. Hellriegel. *Management* (2nd ed.). Reading, MA: Addison-Wesley Publishing Company.

1979

With W. Gifford & R. Bobbitt. Message characteristics and perceptions of environmental uncertainty by organizational decision-makers. *Academy of Management Journal, 22*, 458-481.

With D. Hellriegel. Organizational design: Which way to go? *Business Horizons*, December, 65-76.
With W. Joyce. Climates in organizations. In S. Kerr (Ed.), *Organizational Behavior* (pp. 317-336). Columbus, OH: Grid Publishing Company.
With D. Hellriegel. *Organizational behavior: Contingency views* (2nd ed.). St. Paul, MN: West Publishing Company.

1980

With D. Hellriegel. Preferred organizational designs and problem-solving styles: Interesting companions. *Human Systems Management, 1*, 151-158.
With D. Hellriegel. Assessing organizational change interventions: Towards a comparative typology. *Groups and Organizational Studies, 5*(1), 35-47.
With R. Steckroth & H. Sims. Organizational roles, cognitive roles & problem-solving styles. *Journal of Experiential Learning, 2*, 77-87.
With H. Sims. A typology of technology and job design. *Human Relations, 33*, 193-212.

1981

With S. Kerr. Controlling performances of employees in organizations. In W. Starbuck & P. Nystrom (Eds.), *Handbook of organizations* (pp. 116-134). New York: Free Press.
Job redesign: Improving the quality of working life. *Journal of Experiential Learning & Simulation, 49*, 132-134.
With R. Kerin. Decision-making style and information acquisition. *Psychological Reports, 49*, 132-134.
With P. Lorenzi & H. Sims. Perceived environmental uncertainty: An individual or environmental attribute. *Journal of Management, 7*, 27-43.

1982

With W. Joyce. Correlates of climate discrepancy. *Human Relations, 35*, 951-972.
With K. Downey. Uncertainty and managerial performance. *Social Science Quarterly, 63*, 189-201.
With W. Joyce & M. Von Glinow. Person-situation interaction: Competing models of fit. *Journal of Occupational Behaviour, 3*, 265-280.
Decision making: A psychological perspective. In G. Ungson & D. Braunstein (Eds.), *Decision making: An interdisciplinary inquiry*. Boston: Kent Publishing Company.

With D. Hellriegel. A look at how managers' minds work. *Business Horizons, 26*, 58-68.

With D. Hellriegel. *Organizational behavior* (3rd ed.). St. Paul, MN: West Publishing Company.

With D. Hellriegel. *Management* (3rd ed.). Reading, MA: Addison-Wesley Publishing Company.

1983

Problems with contingency models of leader participation. In J. Hunt & C. Schriesheim (Eds.), *International models of leader behavior* (pp. 333-341). Carbondale, IL: Southern Illinois University Press.

1984

With L. Fry. Structure, technology and workgroup effectiveness: A test of a contingency model. *Academy of Management Journal, 27*, 221-246.

With W. Joyce. Collective climate: Agreement as a basis for defining aggregate climates in organizations. *Academy of Management Journal, 27*, 721-742.

With H. Tosi. Contingency approach: Some suggested revisions. *Journal of Management, 10*, 9-26.

1985

With W. Cron, R. Hansen, & S. Rawlings. Business strategy and the management of the plateaued performer. *Academy of Management Journal, 28*, 133-154.

With W. Cron. Job attitudes and performance during three career stages. *Journal of Vocational Behavior, 26*, 126-145.

1986

With W. Cron. A career stages approach to managing the sales force. *Journal of Business and Industrial Marketing, 1*(1), 51-60.

With W. Cron. The influence of career stages on salespeople's job attitudes, work perceptions, and performance. *Journal of Marketing Research, 23*, 119-130.

With D. Hellriegel & R. Woodman. *Organizational behavior* (4th ed.). St. Paul, MN: West Publishing Company.

With D. Hellriegel. *Management* (4th ed.). Reading, MA: Addison-Wesley Publishing Company.

1987

With W. Cron & L. Yows. Career plateauing: Who's likely to plateau? *Business Horizons, 30*(2), 31-38.

With J. Kerr. Managing corporate cultures through reward systems. *Academy of Management Executive*, pp. 99-108. [Translated, Instituto de Estudios Superiores de La Empresa Universidad de Navarra, Barcelona, 1988. Winner of Best Paper Award, Sponsored by the *Academy of Management Executive*, 1989.]

With J. Kerr. Linking reward systems and corporate cultures. In R. Schuler & S. Youngblood (Eds.), *Readings in personnel management* (pp. 297-308). St. Paul, MN: West Publishing Company.

With S. Stout & W. Cron. Managing superior & subordinate career transitions. *Journal of Vocational Behavior, 30*, 124-137.

With E. Jackofsky. A causal analysis of the impact of job performance on the turnover process. *Journal of Occupational Behaviour, 8*, 263-270.

1988

With E. Jackofsky. CEO roles across cultures. In D. Hambrick (Ed.), *Executive effect: Concepts and methods for studying top managers* (pp. 67-99). Greenwich, CT: JAI Press.

With J. Jolly & T. Reynolds. Application of the means-end theoretic for understanding the cognitive bases of performance appraisal. *Organizational Behavior and Human Decision Process, 41*, 153-180.

With W. Cron. Business strategy and career opportunities. In M. London & E. Mone (Eds.), *H.R. professional and employee career development* (pp. 135-151). Oxford/London: Quorum Praeger Press.

With E. Jackofsky. A longitudinal study of climates. *Journal of Organizational Behavior, 9*, 319-334.

With L. James & W. Joyce. Organizations do not cognize. *Academy of Management Review, 13*, 129-133.

With S. Stout & W. Cron. Dynamics of the career plateauing process. *Journal of Vocational Behavior, 32*, 74-91.

With E. Jackofsky & S.McQuaid. Cultural values and the CEO: Alluring companions? *Academy of Management Executive, 2*, 39-49.

1989

With S. Ornstein & W. Cron. Life stage versus career stage: A comparative test of the theories of Levinson and Super. *Journal of Organizational Behavior, 10*, 117-133.

With D. Hellriegel. *Management* (5th ed.). Reading, MA: Addison-Wesley Publishing Company.

With D. Hellriegel & R. Woodman. *Organizational behavior* (5th ed.). St. Paul, MN: West Publishing Company.

With L. Dlabay. *How to pack your career parachute.* Reading, MA: Addison-Wesley Publishing Company.

1990

With W. Joyce. Strategic context & organizational climate. In B. Schneider (Ed.), *Organizational climate and culture* (pp. 130-150). San Francisco: Jossey-Bass.

With J. Sheridan, R. Buda, & R. Thompson. Effects of corporate sponsorship and departmental power on career tournaments: A study of intraorganizational mobility. *Academy of Management Journal, 33*, 578-602.

With D. Lei & D. Slater. Global strategy and reward systems: The key roles of management development and corporate culture. *Organizational Dynamics*, Autumn, 27-44.

With M. McGill. From inquiry to impact to impotence: The path of consultancy. *Consultation, 9*, 193-198.

1991

With J. Jermier, L. Fry, & J. Gaines. Organizational subculture in a soft bureaucracy: Resistance behind the myth and facade of an official culture. *Organization Science, 2*, 170-194.

With D. Lei. Global strategic alliances, payoffs and pitfalls. *Organizational Dynamics*, pp. 44-62.

1992

With M. McGill & D. Lei,. Management practices in learning organizations. *Organizational Dynamics*, Summer, 5-17.

With W. Cron. Career plateauing. In L.K. Jones (Ed.), *The encyclopedia of career change and work issues.* (pp. 54-56). Phoenix, AZ: Oryx Press.

With C.J. Dilloway. The learning organization. In A. Vicere & V. Freeman (Eds.), *Executive development for global competitiveness* (pp. 45-53). Princeton, NJ: Peterson's Guide, Inc.

With D. Lei. Global strategy, competence-building and strategic alliances. *California Management Review*, Fall, 81-97.

With D. Hellriegel. *Management* (6th ed.). Reading, MA: Addison-Wesley Publishing Company.

With D. Hellriegel & R. Woodman. *Organizational behavior* (6th ed.). St. Paul, MN: West Puiblishing Company.

1993

With M. McGill. Unlearning the organization. *Organizational Dynamics*, Autumn, 67-80.
With D. Lei. Designing global strategic alliances: Integration of cultural & economic factors. In G. Huber (Ed.), *Organizational change, redesign & performance* (pp. 295-322). New York: Oxford University Press.
With W. Cron & E. Jackofsky. Job performance and attitudes of disengagement stage salespeople who are about to retire. *Journal of Personal Selling & Sales Management, 13*, 1-14.

1994

With M. McGill & D. Lei. The new strategy: Anything, anytime, anywhere. *Organizational Dynamics*, Autumn, 33-48.
With M. McGill. Leading learning. *Journal of Leadership Studies*, June, 7-21.
With M. McGill. *The smarter organization: How to build an organization that learns, unlearns, and adapts to capitalize on marketplace needs.* New York: John Wiley & Sons.

1995

With S. Hammond. The impact of prior firm financial performance on subsequent corporate reputation. *Journal of Business Ethics.*
With M. McGill. Executive development in learning organizations. *American Journal of Management Development, 2*, 23-30.
With J. Brett & W. Cron. Economic dependency on work: A moderator of the relationship between organizational commitment and performance. *Academy of Management Journal, 38*, 261-271.
With J. Sheridan & R. Buda. Factors influencing the probability of employee promotions: A comparative analysis of human capital, organization screening and gender/race discrimination theories. *Journal of Business and Psychology, 11*(3).
With D. Hellriegel & R. Woodman. *Organizational behavior* (7th ed.). St. Paul, MN: West Publishing Company.

NAME INDEX